Alternative Home Heating

Dan Browne

Alternative Home Heating

Holt, Rinehart and Winston New York

Published by Holt, Rinehart and Winston, 383 Madison Avenue, New
York, New York 10017.
Published simultaneously in Canada by Holt, Rinehart and Winston of
Canada, Limited.

Library of Congress Cataloging in Publication Data
Browne, Dan.
Alternative home heating.
Includes index.
1. Dwellings—Heating and ventilation.
2. Dwellings—Energy conservation.   3. Solar heating.
4. Wood as fuel.   5. Heat pumps.   I. Title.
TH7226.B75   697   79-3445
ISBN Hardbound: 0-03-041531-4
ISBN Paperback: 0-03-041536-5

All photographs were taken by Daniel B. Doman.
Diagrams are by Kevin McCloskey.

FIRST EDITION

Designer: Joy Chu
Printed in the United States of America
10 9 8 7 6 5 4 3 2 1

FOR LISA

Contents

Alternative Home Heating

Introduction

Over the past ten years or so there has been a determined search for alternatives to the increasingly expensive use of oil and gas, and of electricity generated by fossil fuels. Renewable-energy schemes still in the conceptual stage include harnessing tides and tapping the earth's magma for heat, and there are others, such as nuclear fusion, that are in the earliest laboratory stages. Experiments are underway for obtaining energy from aquatic crops, burning coal underground in controlled fires, and storing electricity at near-absolute zero. Facilities that generate electricity from natural hot springs continue to be built, though their impact on the energy scene is slight. Photovoltaic cells that convert sunlight directly into electricity have been demonstrated to be feasible but are simply too expensive to use except in isolated instances. Wind as a source for generating electrical power has also proved very expensive with only limited application.

All these possibilities, attractive as they may seem, are still remote from our daily lives, from those systems, devices, and practices that can be used today in our country's sixty million single-family homes. Only limited solar-energy systems, heat pumps, and wood-burning devices offer realistic and immediate relief from the increasingly burdensome cost of filling our home-energy requirements.

1

I have yet to meet anyone who didn't agree that the wholesale use of solar energy as a replacement for fossil fuels is much to be desired. In 1974, reflecting this widely held view, Congress enacted the Solar Heating and Cooling Demonstration Act calling for "programs to develop, demonstrate, and promote the use of Solar Heating and Cooling systems." During 1978 alone, the Department of Energy funded more than five thousand solar projects. Companies have been formed to manufacture solar components, a national retail chain has set up a solar-energy division geared to residential use, and periodicals devoted exclusively to solar energy have flourished. But in spite of the thrust provided by publicly funded projects and a great deal of favorable publicity, solar energy has failed to catch hold.

That failure has been evidenced recently in Connecticut. State legislation passed in 1977 grants $400 to the homeowner who installs solar energy, and although seven hundred grants were available, only two hundred homeowners applied in 1978. And virtually all applicants intended to install a hot-water system rather than a space-heating facility. (Space heating consumes the bulk of the energy used in a home.)

Such consumer hesitation exists throughout the country. Solar activity in the United States is essentially limited to hot-water supply in the Southwest and the lower half of Florida. In 1977, out of a total of some four million square feet of solar collectors manufactured, more than three million were used to heat swimming pool water. Here again, these installations were mostly relegated to the same two regions of superabundant sunshine. It appears that while homeowners are willing to acknowledge solar space heating as a first-rate idea, they are not prepared to install systems in their own homes.

Why has the response been so poor, particularly with all the editorials telling us that solar energy will not pollute the atmosphere, will preserve our nonrenewable fossil fuels, will lessen our dependence on foreign oil and thereby improve our trade balance, will reduce inflation and even save us money?

A glance at the history of solar energy partially explains why these powerful arguments have not persuaded us.

The significant use of solar energy dates from the 1920s, when some sixty thousand solar systems in Florida supplied household hot water. It was a remarkably high usage (in contrast, there are only forty thousand solar installations in all of the United States today), especially when one considers that hot-water supply by conventional means then cost less than a dollar a month and installing a solar system cost several hundred. To make solar use more attractive, the system was advertised as "free hot water for a lifetime." Simultaneously, a widespread practice developed of introducing cheaper components to lower the initial costs. For the first few years of such a system, hot water was supplied abundantly, but then corrosion and other problems due to the inferior components caused malfunctions that were neither easy nor cheap to correct. Homeowners felt they had been cheated by the misleading advertising, and many who had rushed to install solar systems had them removed and went back to conventional water heaters. Builders began to advise prospective homeowners against solar use.

Solar energy, by then viewed with suspicion, remained dormant until the mid-1940s. In the explosion of housebuilding that followed World War II, scientists associated with the Truman administration spearheaded a renewal of interest in solar heating, and homeowners throughout the country were exposed for the first time to government proclamations of "the energy future." These scientists, ignoring what had happened in Florida, extolled the virtues of solar energy extravagantly and, carried away by their own enthusiasm, predicted fifteen million solar homes by 1978. Energy was still cheap, however, and conventional water heaters continued to be used in all but a few instances. Solar space heating fared even worse.

The situation remained essentially unchanged until the energy crisis of the mid-seventies, when once again solar en-

3

ergy burst upon the scene and became a household word. A goal of 2.5 million solar homes by 1985 was proclaimed, this time by the Carter administration. False and misleading assertions continue throughout solar literature, and even the Department of Energy, the new, primary force for development in the field, has not been immune. Until 1978, DOE literature repeatedly stated that the efficiency of solar systems was 50 percent (percent used of available solar energy), even after New England Electric, in the first solar domestic-hot-water program of significant size (one hundred single-family homes were equipped with solar systems and monitored), found average efficiency to be 17 percent ("Interim Report on the New England Residential Solar Heating Experiment," New England Electric, Westborough, Mass., 01581, 1977). Such fundamental misrepresentations continue to plague the field, much to the detriment of real progress, and show that the mentality of "free hot water for a lifetime" is still with us.

In this confused and confusing situation, the homeowner can best distinguish between fact and fiction by examining a solar system very closely—how it performs in practice, what it will cost to install, and what benefits, if any, will be derived.

These are the practical considerations I have addressed myself to through the first section of this book, which covers the principal alternative systems of solar heating: solar space heating (liquid medium), solar hot water, solar hot water for pool and household, solar window and insulated shutter, passive solar concrete wall, passive solar modular wall collector, and solar greenhouse.

I begin with solar space heating as a means of introducing the components and operation of typical solar systems, although I will demonstrate why I consider solar space heating through a liquid medium to be economically unsound and fundamentally flawed for efficient home use. All the other systems have merit in varying degrees.

The book's second section is devoted to fireplaces and

stoves with wood as an alternative fuel. The notion of return-
ing to the widespread use of wood may seem far-fetched at
first, but many inefficiencies and inconveniences associated
with wood heating can now be easily and cheaply eliminated,
and there are enough woodlands to fill most all our residential
heating needs.

Wood is our one major natural resource that has become
more abundant in the last century. In populous regions such as
the Northeast, millions of acres of former crop lands have re-
verted to their original state of woodlands. Similar transfor-
mations have occurred in other regions, swelling the firewood
supply and bringing it closer to major population centers. Today that supply is so abundant and accessible, it would renew
itself even if all sixty million homeowners burned wood exclusively!

The price of firewood has fluctuated wildly in the last
year or so, but the recent increases in populous areas of the
country have been due more to dealers capitalizing on the
wood-burning boom than to actual supply and demand. In
the Catskills, where I live, I still pay $60 per cord of dry oak,
and the price next year is likely to go *down*. For homeowners
in urban and suburban areas where firewood profiteering is
most prevalent, a simple remedy is for four or five households
to get together and order sixteen to twenty cords (one truck-
load) to obtain the lower prices that prevail in areas where
wood supplies are cheap and plentiful.

Homes in climates where snow covers the ground between November and April generally need four to five cords
per year, and although wood is often burned inefficiently, the
cost is already competitive with running an oil furnace.

The economic picture for firewood will undoubtedly improve. Further price decreases are likely as suppliers become
more mechanized and their operations more efficient. In many
areas one also has the option of cutting firewood without
charge in state forests or culling nearby woodlands for the
firewood. Woodlands mature well without human help. No

large capital expenditures are involved, and labor is limited to gathering and delivering. The trees themselves are far more efficient and cost-effective than any other means of collecting and storing solar energy.

Wood is a fuel that does not pollute the atmosphere. When it is burned the process of photosynthesis is simply reversed. Carbon dioxide is the main ingredient of the exhaust, and an equal amount would be released to the atmosphere if the wood were to rot where it had fallen.

Although wood as a fuel is environmentally and economically sound, there are, of course, disadvantages to its use. The most obvious is the need to fetch the wood from its place of storage, feed the fire, and remove the ashes. There can be other disadvantages as well, depending on whether wood is used exclusively or partially.

If wood is used exclusively, water pipes have to be drained when one leaves for a weekend with freezing temperatures anticipated. Kerosene has to be dumped in the toilet bowl and antifreeze cycled to the pump of the washing machine. At these and other times, without the convenience of thermostatically controlled automatic heat, periods of discomfort can be expected. (Of course, there will also be the advantages of eliminating the depreciation and cost of running and maintaining the furnace.)

If wood is used partially, in conjunction with a furnace, the advantages of both are retained. With the furnace used only when thermostatically controlled heat is needed, and wood at all other times, comfort and convenience are obtained in a highly cost-effective way.

Both arrangements—exclusive and partial—are valid under different circumstances, and the homeowner has to decide which is better suited to the specific situation. But no matter which choice is made, one will naturally want to burn the wood as efficiently as possible, and the section on fireplaces and stoves is directed toward this end.

Fireplaces in particular have long been known to be very

*in*efficient, and although there have been significant improvements in heat recovery, these have been restricted to rare and isolated installations. Virtually all modern fireplaces are essentially the same as those built in colonial times; they still lose 90 percent of their heat up the flue.

At a cost that is amortized in a year or two of use, the efficiency of an existing fireplace can be dramatically improved without altering its structure. At a cost that is amortized over seven to eight years, the structure of an existing fireplace can be altered to obtain a higher efficiency than an oil furnace and to provide heat more cost-effectively. Such detailed improvements are presented in the fireplace chapter.

Like fireplaces, modern stoves are essentially the same as their predecessors of a century ago, and some of the same manufacturing equipment used then has now been dusted off and put back to work. Stove designs are based on ease of fabrication and cosmetics; neither has much to do with getting a maximum amount of heat into the room. The chapter dealing with stoves is mainly concerned with achieving such high efficiency and presents detailed means of doing so.

The final section deals with the heat pump, an old concept returned in modern form. Heat pumps have been reintroduced with a good deal of publicity and, as might be expected, with false and misleading assertions. While there are situations in which the heat pump is an optimum choice, one should approach these machines keeping in mind that they are powered by electricity and that utility companies have regularly deceived homeowners.

During the 1950s and 1960s, for example, utility companies mounted a large-scale advertising campaign that presented electricity as "clean, 100 percent efficient heat." While it is true that electricity used to heat a home does not pollute and is completely converted to heat, both parts of the statement are misleading. The oil burned by the utility to generate the electricity polluted the atmosphere, and 65 percent or more of the energy in the oil was lost in the generation and transmission of the electricity. By the time electricity is used in a home, its efficiency is already down to 35 percent. Of course, the homeowner pays for this *least efficient form of heating*. Twenty years ago a great many homeowners were taken in by this deception and went "all electric." We were then still in the cheap-energy era and the impact was softened, but electrically heated homes today are a heavy and sometimes catastrophic burden to owners.

A similar deception is currently being presented about the heat pump. Certain manufacturers' television commercials have told viewers that the heat pump is "two and a half times more efficient than 100 percent efficient electric heat." What we are not told is that this applies only when the outdoor temperature is 47 degrees, the testing standard, and that *heat-pump efficiency drops markedly as the outdoor temperature decreases*. Since most of us live in climates where lower winter temperatures are the rule, the assertion is grossly misleading. And of course the deception inherent in "100 percent efficient electric heat" is perpetuated.

The material in the final section shows what a heat pump is, how it works, where and in what forms it is suitable, how much it will cost to install, and what benefits will be obtained.

Whether motivated by economic or environmental considerations, or both, an increasing number of homeowners feel that their present conventional methods of providing heat and hot water are getting out of hand and that change is necessary. Unfortunately, no single alternative fuel or heating system is appropriate to all situations. If this book provides the basis for selecting one's own optimum choice, a choice that is both cost-effective and nonpolluting, its purpose will have been fulfilled.

SECTION ONE
Solar Heat

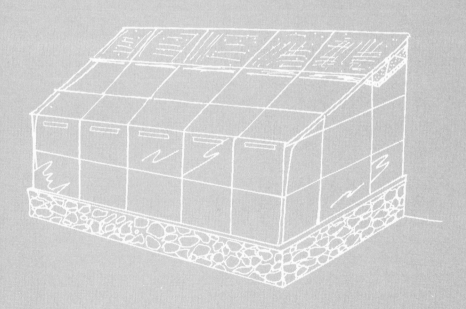

Solar Space Heating (Liquid Medium)

The rectangular objects in Figure 1 are *solar collectors*, and the diagram (Figure 2) shows them as components of a solar space-heating system that serves a recently built addition to a house in Connecticut. A total of sixteen such 4-by-8-foot collectors perform the same function that a furnace would in a conventional system and provide heat for a 1,500-square-foot building. The collectors, a flat-plate variety, and the system of which they are a part, a liquid-medium type, are both typical; systems such as this one account for 80 percent of all solar space-heating installations.

The collectors were placed on the roof of the building for convenience and to avoid shade. They face true south to obtain the most effective sunlight each day, and they are inclined at a 56-degree angle for optimum winter use.

Instead of burning fuel, collectors obtain heat directly from the sun. Several minutes after radiation is produced on the sun's surface, it reaches earth in the form of parallel electromagnetic waves of different lengths. The ultraviolet portion of the radiation provides energy for plant growth; the infrared portion heats the earth.

11

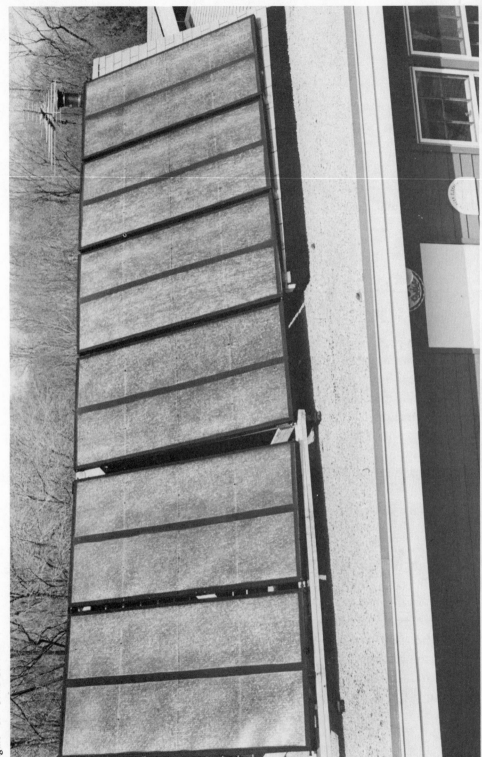

Figure 1. Solar Collectors

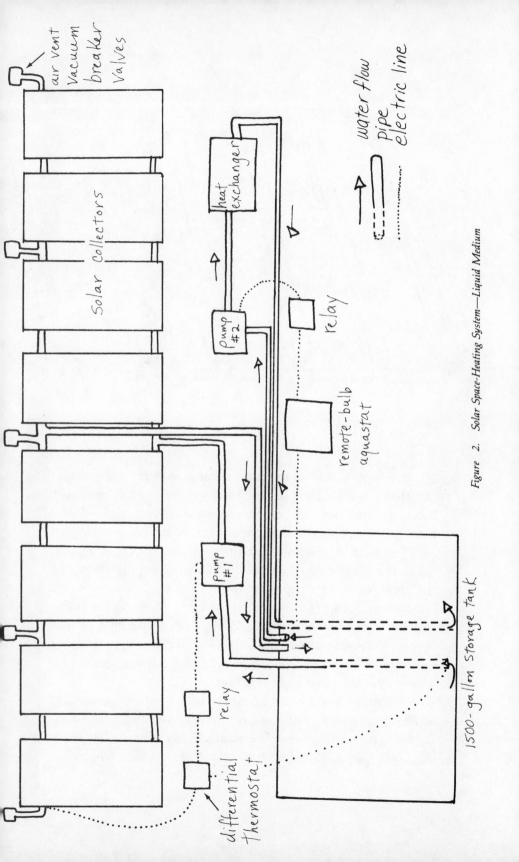

Figure 2. Solar Space-Heating System—Liquid Medium

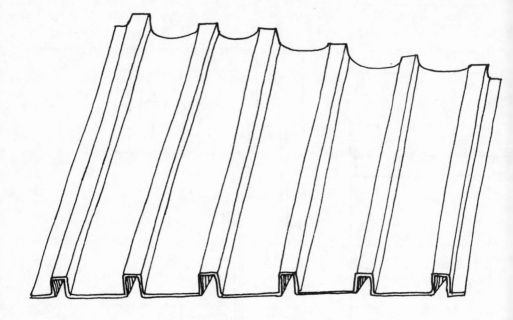

Figure 3. Absorber Plate

The way collectors function is based in part on the simple fact that when sunlight strikes an object cooler than itself, the object absorbs heat. Collectors are essentially specialized surfaces for absorbing a maximum amount of heat from sunlight.

The part of the collector that absorbs the sun's heat is called the *absorber plate* (Figure 3). A sheet of copper about the thickness of heavy paper, it is located an inch below the translucent cover, from which it is separated by an air space. The U-shaped indentations make the plate more rigid so that it doesn't develop undulations that would cause the sun's rays to strike its surface at a variety of angles (absorption is best when rays strike at a right angle).

As sunlight shines on a bare sheet of copper, a good deal of heat is absorbed, but some is also reflected and lost. To reduce the amount reflected, a matte finish of a bluish-black metallic oxide is coated on the surface of the plate. The selective coating reduces *reflectance* to an insignificant amount.

Some of the heat the plate amasses is simultaneously lost to *emissivity*—the outward passage of heat from the surface of the plate to cooler surrounding air—but the selective coating reduces such losses to a very small amount.

Unenclosed, the plate would be cooled by breezes in the same way a hot liquid is cooled by blowing air over it, an action called *convection*. Even on a still day, convective losses would occur as heat emitted from the plate encountered cooler surrounding air; the warmed air would rise and be replaced constantly by cooler air to absorb the emitted heat. To prevent convective losses, the plate must be enclosed. Figure 4 shows the *collector case*, which partially encloses the plate. It is made of aluminum and painted black to increase its functioning temperature and thereby minimize heat loss.

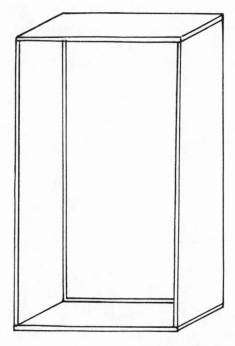

Figure 4. Collector Case

Front and back covers are needed to create an airtight container completely enclosing the plate. The front, facing the sun, is itself called the *cover* and must not keep the sun's rays from striking the plate. The cover shown in Figure 5 is a sheet of translucent fiberglass that is impact-resistant and nondiscoloring. It is attached to the underside of the case perimeter by double-faced, pressure-sensitive adhesive strips and a bead of neoprene.

In addition to its other functions, the cover serves to trap heat produced inside the collector. Figure 6 shows the short waves of the sun's radiation passing through the cover, striking the plate, and reradiating as long waves. Short waves have no difficulty penetrating the fiberglass; but once converted to long waves, they no longer pass through the cover as easily. Long waves accumulate between the back of the cover and front of the plate and give rise to a superheated condition known as the *greenhouse effect*.

The distance between the cover and plate is an important factor in making full use of the greenhouse effect. As the gap between the two is decreased, there is a greater concentration

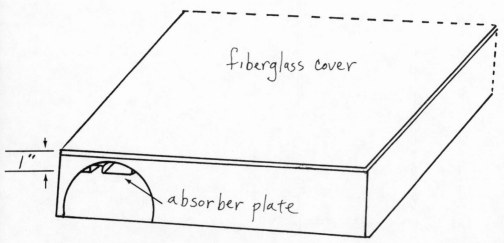

Figure 5. Collector Cover

16

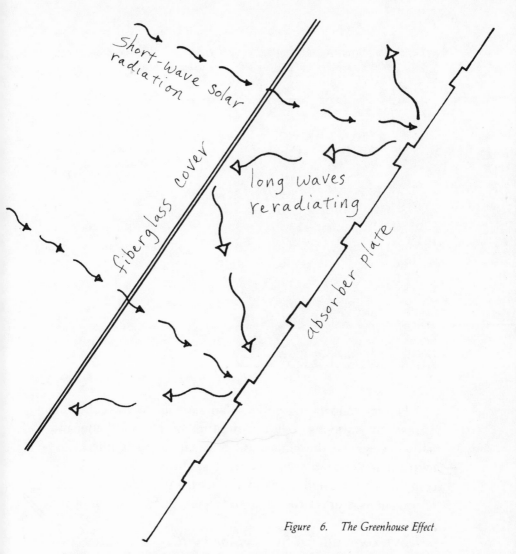

Short-wave solar radiation

Fiberglass cover

long waves reradiating

absorber plate

Figure 6. The Greenhouse Effect

of long waves and a greater heat loss by *conduction* through the cover. As the distance between the two is increased, greater air turbulence results and causes greater heat loss by convection. Some heat loss is unavoidable, but studies have shown that the smallest loss occurs when the air gap is 1 inch.

To maintain the desired gap, 1-inch metal spacers are attached at right angles to the plate indentations and cover. The spacers also provide additional strength to the assembly and help keep the cover and plate in a flat plane.

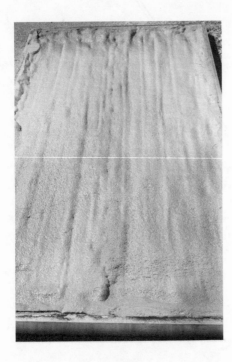

Figure 7. Polyurethane Backing of Collector

Figure 7 shows the polyurethane *backing* of the collector, formed by spraying a liquid mixture of polyurethane and hardener against the plate until the rear cavity is filled. The mixture hardens rapidly into a rigid, slightly spongy mass that completes the airtight container and also provides the necessary insulation and structural reinforcement for the rear of the plate.

In a conventional furnace, the burning of fuel heats air which is then delivered through the house. The action is called *heat transfer*, and in this instance the *transfer medium* is air. In an electric water heater, electricity is conducted through a nichrome element, and resistance to its flow creates heat. The heat is then transferred to the surrounding water—the transfer medium in this system—and delivered through the plumbing. To remove the heat amassed by solar collectors, one may use a *gas medium*—usually air—or more likely, a *liquid medium*—usually water or a mixture of water and antifreeze.

The collector's *heat-transfer* component consists of ¼-inch copper pipes called *transfer tubes* that are positioned in the indentations of the plate and brazed at both ends in holes of ½-inch copper pipes called *manifolds* (see Figure 8). Water was used as the medium of this installation because it is more efficient than air and less expensive than a mixture of water and antifreeze.

Water is delivered through pipes to the lower manifolds and rises into the transfer tubes. As the water proceeds upward, heat from the plate is conducted to it through the tube walls. The heated water then flows into the upper manifolds, spills into a return pipe, and runs back to its source. (The ends

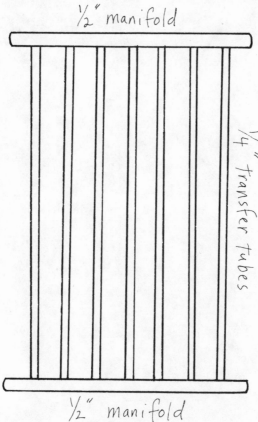

Figure 8. Heat-Transfer Component (2-Foot Section)

of the manifolds extend outside the case so that collectors can be joined, with inlets and outlets provided for each.)

Figure 9. Insulated Solar Storage Tank

The heat amassed by the collectors and transferred to the circulating water can be used immediately. However, when an excess is produced or there is no demand, a means of storage is needed. After all, heat must be available for use at night or whenever the sun isn't shining. (During winter in Connecticut, for example, the collectors have no source of heat 80 percent of the time.) Figure 9 shows the insulated 1,500-gallon tank that is used to store solar-heated water. The *storage tank* is covered with 2-inch-thick Styrofoam insulating panels and is waterproofed by neoprene.

A garden hose inserted through a manhole at the top is used to fill the tank. As the water is circulated and heated, it separates naturally into layers (*stratifies*) and the coldest portion, being heaviest, lies at the bottom. A 1¼-inch copper pipe with an open end is installed near the bottom of the tank so that this coldest water can be withdrawn first at the start of each cycle and directed toward the collectors.

Figure 10.   Pump #1

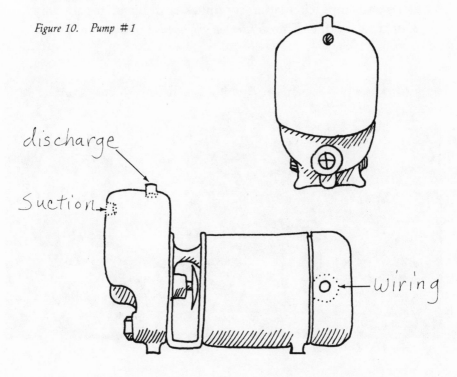

Figure 10 shows *pump #1*, which circulates water from the tank through the collectors and back to the tank. It is a centrifugal type that employs a fractional horsepower electric motor and is powerful enough to raise water to the upper manifolds 17 feet above. Using standard data, the optimum rate of flow was calculated to be 6 gallons per minute.

In a conventional heating system one merely sets the desired temperature on a thermostat and the flow of heat is regulated. In the solar situation we are not concerned with *absolute* temperature but with the *difference* in temperature between the absorber plates and tank water. There is no point in circulating water at night, for example, since the collectors aren't amassing heat; frequently they may be colder than the water, so heat from the water would actually be lost. Experience has demonstrated that the best time to begin circulation is when the absorber plates are 20 degrees hotter than the tank water (all temperatures are Fahrenheit). In order to avoid frequent starts and stops, circulation continues until the plates are only 3 degrees hotter. The *differential thermostat* shown in Figure

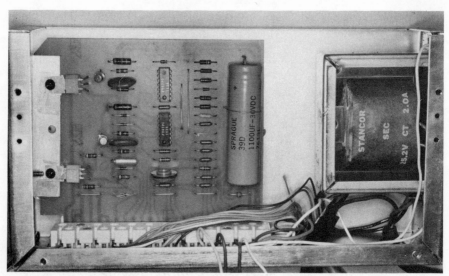

Figure 11.   Differential Thermostat

Figure 12. Thermistor Sensor

11 controls the circulation of water between the tank and collectors; as its name implies, it operates on the difference in temperature between the two.

The differential thermostat needs to "know" the temperatures of both the water and the plates before it can act. The *thermistor sensor*, at one corner of the collector plate (Figure 12), does this. The sensor is a thin, temperature-sensitive wire that acts much like a thermometer. One sensor is attached to a plate inside a collector and another near the bottom of the tank. The sensors read the temperatures at both locations and transmit the information to the differential thermostat via attached wires.

The differential thermostat operates on 24 volts and does not directly start or stop the pump (which runs on 120 volts) but works in conjunction with the relay shown in Figure 13. The relay is installed in the electrical path between the ther-

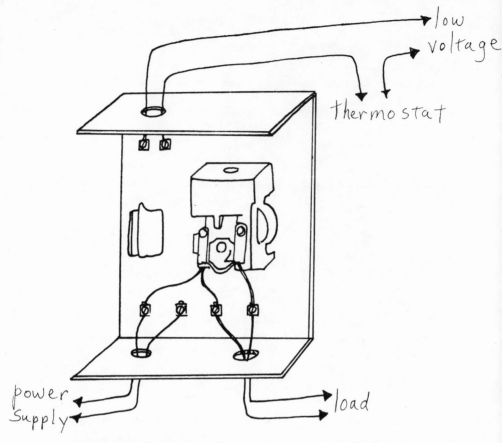

voltage

thermostat

power
Supply

load

Figure 13.   Thermostat-Pump Relay

mostat and pump motor. When it receives an impulse from
the thermostat, it opens or closes its contacts to stop or start
the pump motor.

As we have seen, pump #1 lifts water from the storage
tank to the collectors. When the pump stops, the pressure
ceases and the system begins to drain. Water in the upper
manifolds falls through the return pipe and is deposited at the
top of the tank. The remaining water drains back through the
transfer tubes, reversing its former path. The receding water
leaves space behind, and unless air is admitted to fill the
space, a vacuum develops and prevents the water from draining. (In winter, of course, any remaining water would freeze
and burst the transfer tubes, a common occurrence.) Figure 14

Figure 14.   Air-Vent Vacuum-Breaker Valve

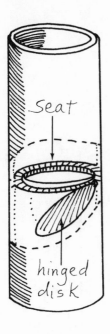

shows an *air-vent vacuum-breaker valve* that automatically permits air to enter behind the draining water. The valve is equipped with a hinged disk inside the housing. During circulation, water pressure forces the disk upward against a stationary seat and closes the outlet. When the pump stops and pressure ceases, the disk falls away from the seat, exposing a $\frac{1}{2}$-inch hole that allows air to rush in to fill the space left by the receding water. Five such valves are spaced along each 32-foot array.

The system so far described gathers and stores the sun's heat. Solar systems do not supply all the heat required by a structure; a goal of 70 percent has been widely accepted as the maximum practical objective. It follows that at least 30 percent of the heat will have to be supplied by an alternate source. In this instance, a conventional oil-fired furnace was installed for the purpose. The heat of both sources is delivered through conventional hot-air ductwork.

Figure 15, a *heat exchanger*, is the component that transfers heat from the tank water to the air. Inside the exchanger there is a coiled copper tube with many closely spaced metal fins

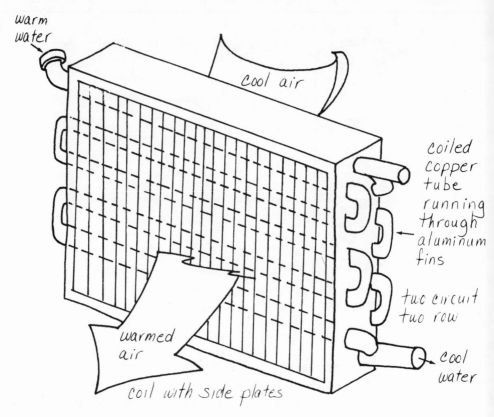

Figure 15. Liquid-to-Air Heat Exchanger

attached to its outer walls. The exchanger is located inside the main duct of the furnace. When there is a demand for heat, hot tank water begins to circulate inside the coil and heat is rapidly conducted through the walls to the fins. Simultaneously, the furnace fan draws a large volume of relatively cooler air across the fins and the heat is transferred to the air. The warmed air is then delivered in the conventional way.

Pump #2 (Figure 16) is used to circulate tank water through the exchanger. It is similar to pump #1 but less powerful, since it is designed to lift water only 6 feet rather than 17 feet.

Temperature is the driving force of heat, and in any heat-transfer operation it is best to deliver the medium surrendering heat at the highest temperature and the medium receiving heat at the lowest. The inlet pipe of the pump is therefore po-

Figure 16.   Pump #2

Figure 17.   Bulb:
Remote-Bulb Aquastat

sitioned near the top of the tank, where water is hottest, and the return empties near the bottom.

A *remote-bulb aquastat*, an electrical device similar to a conventional thermostat, is used to control the flow of water from the tank to the heat exchanger. The *bulb* (Figure 17), the sensor part of the device, lies a few inches below the surface of the tank water and reads the temperature of the hottest layer. The reading is transmitted to the aquastat, which in turn starts or stops pump #2 via a relay. When the tank water is 80 degrees or higher and the thermostat demands heat, circulation begins automatically and continues until the demand is supplied. If a demand for heat occurs and the water is less than 80 degrees, the aquastat prevents circulation and the solar system remains off. (In this installation, the oil-fueled backup furnace is then turned on.) If the water is 80 degrees or higher at the start and falls below 80 before the demand for heat is satisfied, the aquastat shuts off circulation.

A two-stage thermostat regulates the flow of heat to the structure. As with an ordinary thermostat, which it resembles, the device is simply set at the desired temperature and the rest is automatic.

The first stage of the thermostat is wired to the solar heating system, the second stage to the backup furnace. When heat is needed, the first stage turns on the pump and fan motors. Hot water circulates through the exchanger as air is drawn over the fins and the heated air is ducted to the rooms. Once the desired temperature is reached, the thermostat shuts down the system.

But if heat is needed when the tank water temperature is below 80 degrees, the aquastat prevents circulation. The room temperature continues to fall, and at a preset level—a couple of degrees lower than the setting for the first stage—the thermostat activates the second stage. The furnace comes on and heat is provided in the conventional way until the demand is satisfied.

Performance

A highly energy-efficient structure loses approximately 20 btus per hour per square foot of area when the outdoor temperature is zero degrees and the indoor temperature 70 degrees. (Btu, British thermal unit, is the standard of heat measurement; 1 btu is the amount of heat required to raise 1 pound of water 1 degree—a kitchen match yields approximately 1 btu.) The product of this figure for any one structure is known as that structure's *design heat loss* and provides the basis from which various space-heating needs are determined. The design heat loss of the addition we have been describing is 30,000 btus, representing the amount of heat it requires to replenish the hourly loss under the above conditions.

During the early part of the heating season when ambient temperatures range from 50 to 65 degrees, only a few thousand btus are needed to maintain a 70-degree interior temperature. Tank water temperatures, 140 degrees and higher during summer, are maintained well above 110 degrees (the minimum required for efficient operation of the exchanger) and heat transfer to air is effective. Even after a couple of partially overcast days there is enough heat stored in the tank water to fill the needs of the addition; the furnace, therefore, is seldom used during this favorable period.

Although there is relatively little demand for heat, average tank water temperature begins to fall on account of heat seepage from the tank and piping, despite the insulation covering them. Colder and stronger winds blowing across the collectors reduce heat input and contribute to lower tank water temperature. The losses are substantial but not too apparent, because more heat than the structure needs is still generally available.

As the cooler months progress and temperatures dip into the 40s, the demand for heat increases correspondingly, and excess stored heat is rapidly consumed. Heat loss from seepage increases and begins to play a major role in lowering efficiency.

During the period of late fall and early winter, the availability of solar energy declines and less heat is amassed. The combination of lower solar input, heat seepage losses, and greater demand reverses the earlier situation; the backup furnace is now needed to supply about 65 percent of the heating demand.

As these changes occur, the average temperature of the tank water continues to decline; it is often less than 100 degrees. Even after a sunny day the net temperature gain is only a few degrees, since most of the amassed heat is either used or lost. During partially overcast days there is a net loss of several degrees due mainly to heat seepage. Since roughly two-thirds of the days during late fall and early winter are partially or completely overcast, the unfavorable trend of lower tank water temperatures is dominant. Temperatures below 110 degrees seriously impair the efficiency of the heat exchanger, and the furnace is used much more often. A substantial part of the amassed heat stored in 90-degree water, for example, is never exchanged; it is simply lost through seepage from the system.

In midwinter—when ambient temperatures are in the 30s and below, and heating needs are greatest—solar energy is least available, and heat losses are at their maximum. Tank water temperatures are often below 80 degrees, so no heat exchange is possible. During a sunny day, most of the heat transferred from the collectors is used to raise stored water to the minimum operating temperature and only a minor amount is actually used to heat the building. The situation improves following consecutive days of sunshine, but the gain is canceled by twice as many sunless stretches. *The average amount of heat utilized in midwinter is rarely more than 10 percent of the total required to keep the building at 70 degrees, and often less.*

We can see why this occurs by examining the solar heating process more closely. Solar radiation falls on the collectors in a New England December at an average daily rate of 1,333 btus per square foot. The collectors are about 40 percent efficient and absorb 533 btus per square foot of the available radiation. An additional 20 percent or so of the absorbed heat is lost during transfer and storage, leaving a daily net of 426 btus per square foot. Since the collectors have a total of 450 square feet of absorptive area, the net daily input into the system totals 191,700 btus.

Let us assume that the tank water temperature is 70 degrees at the start of an average day. Before heat transfer can begin, 122,495 btus are needed to elevate the tank water to the minimum operating level of 80 degrees (8.33, the number of pounds per gallon, times 1,500, the number of gallons, times 10, the number of degrees). It is therefore apparent that the majority of the heat amassed during an average day is used to heat the tank water and not the building.

After the water has been brought up to 80 degrees, 69,205 btus (the equivalent of a 5.5-degree temperature rise) still remain from the total amassed and are theoretically available for heating the addition. However, when supplied with 80-degree water, the heat exchanger has extremely poor efficiency and only a very small amount of heat is transferred during each cycle. The water continues to circulate at each demand for heat and its temperature *rises* as additional btus enter the system and a lesser amount is extracted or lost. The rise in temperature is both small and slow; usually it reaches a maximum of 84 degrees. At this point only 19,225 btus remain of the amassed heat; the rest has been lost or used in elevating the tank water temperature from 70 to 84 degrees.

The efficiency of the heat exchanger is improved slightly when supplied with water of 84 degrees rather than 80 degrees, but it is still poor; only very small amounts of heat are exchanged during each cycle. There is also a small amount of heat lost due to seepage during each cycle, at a rate higher than if the water were merely stored in the tank. As ambient temperatures fall after sunset and the demand for heat increases, the stored water is circulated even more frequently, and greater seepage losses occur. As the tank water temperature declines further, still less heat is exchanged and losses continue to mount. The process usually continues for several hours after sunset before the tank water falls below 80 degrees and the system shuts itself off. For the remainder of the night, water stored in the tank loses heat at the average rate of 0.7 degrees per hour. By next morning the water is usually about 72 degrees.

During this twenty-four-hour period, the bulk of the heat amassed by the collectors was used to elevate the tank water to 84 degrees. Of the 19,000 or so btus that remained, approximately 13,000 were actually delivered to the building, with the balance lost by seepage during circulation and storage. But 13,000 btus are most often less than 10 percent of the total heat needed during a December day.

The poor winter performance of this system is not unusual. A similar space-heating system was constructed and monitored for two years by an MIT research team. Their report states that average solar incidence for December was 40,730 btus per square foot monthly, or 444 btus per square foot daily. The overall system efficiency worked out to be 33 percent. If we apply the same net figure to the Connecticut system, the total input per day would be 199,831 btus—within 5 percent of the actual result in Connecticut.

The solar system supplies approximately 20 percent of the total heat used by the addition over the course of an entire heating season. The bulk of the heat is provided during early fall and spring, a much smaller amount during late fall and spring, and a minor amount during winter. Poor winter performance is not attributable to the heat exchanger, a highly efficient component with a proven record of performance over many years in a multitude of other heat-transfer operations. The crux of the problem is that 1,500 gallons of water is simply too great an amount for the collectors to maintain at the 110-degree temperature required for efficient heat transfer.

The system was designed to last at least twenty years. All components are of good quality and were purchased at professional discount prices, with the savings passed on to the owner. A good deal of the labor of design, assembly, and erection was donated and does not appear in the total cost; there-

Table 1
COSTS OF SOLAR SYSTEM
(1979 prices used throughout)

Collectors	$5,248
Framework	819
Pumps	417
Controls	542
Tank and Waterproofing	1,216
Heat Exchanger	333
Plumbing—Labor and Materials	320
Insulation, Electrical, and Miscellaneous	256
Utility Room for Solar Only	1,536
Professional Services	1,024
Hot Water—Labor and Materials	512
Total Cost of System	$12,223

fore, the total initial outlay is well below average. It should be pointed out, however, that a domestic hot-water supply, not dealt with in this chapter, is invariably part of solar space-heating systems, and the additional benefits and liabilities derived, namely cost of hot-water supply, *are* included in the above table.

The cost of running and maintaining the system must also be included before an economic conclusion is drawn. One can reasonably estimate that each pump runs 1,000 hours yearly and the furnace fan 300 hours longer than required in a conventional system, because the lower temperature of the heat source in a solar system requires longer air circulation to satisfy heating demands. The cost of running the three motors plus a few other incidentals amounts to $70 yearly.

The exact outlay for maintenance is difficult to predict, but experience provides the basis for a reasonable estimate. The pumps, although of superior quality, will not last twenty years, based on the performance of the same pumps in similar situations. One can expect a loss of seal, impeller fracture, or other malfunctions in ten years or so. Similarly, it is unrealistic to expect the electrical components to last twenty years. The relays and fan capacitor, in particular, are very likely to malfunction within seven years. The malfunctions expected of these and other components are not due to poor quality but reflect "the state of the art." All active mechanical systems require maintenance, and solar systems are no different in this respect. A reasonable cost estimate for average yearly maintenance is $100, an amount close to that of the much simpler backup system.

Total operating costs are therefore $170 yearly.

The value of fuel saved during the first year of operation was $398. If the cost of operation and maintenance is subtracted, the balance of $228 is the dollar equivalent of the fuel saved during the first year.

A fuel savings of $228 is obtained assuming that all the solar energy produced is used. But of course, during summer months, when there is no need of space heating in many homes, more heat will be produced than is needed, and unless a way is found to use it, the excess has to be discarded. The excess was calculated to have a yearly value of $169. If this amount is subtracted, the first year of operation will net $59. The homeowner who cannot use the heat produced in summer is faced with the situation of trying to amortize an initial outlay of $12,223 with a $59 payment at the end of the first year. Obviously this is impossible; the projected twenty-year life-span will end long before the initial outlay can be repaid.

The uses of excess solar energy are limited, but let us assume, as is true in Connecticut, that it can be used to heat swimming pool water. Then the value of fuel saved is the full $228. If the savings of the first and all subsequent years are applied toward amortization, the following picture emerges.

Table 2 shows that if a homeowner can use all of the excess summer heat, the initial outlay will be repaid between the nineteenth year and twentieth year. However, only a half year or so remains of the system's twenty-year life-span before complete loss of the original investment. If $12,223 is

Table 2

Years	Annual Fuel Savings *	Cumulative Total
1	$228	$228
2	251	479
3	276	755
6	367	1,059
10	537	3,633
17	1,088	9,243
18	1,153	10,396
19	1,268	11,664
20	1,395	13,059

* A 10 percent yearly increase in the price of fuel is allowed.

deposited in, say, a savings account, and the earned interest partially applied to increased fuel costs, the capital will not only be intact after twenty years, but will have risen considerably.

If one considers space heating apart from hot-water supply, yearly fuel savings are only slightly higher than the cost of operating and maintaining the system, and thus the economic picture is even bleaker. Unless the initial outlay can be reduced by more than half and heat-transfer efficiency raised dramatically—changes not yet on the consumer horizon—one can only conclude that a solar space-heating system in temperate zones that experience freezing winters, even with a backup furnace, makes no economic sense and should be avoided by the homeowner.

Solar Hot-Water System

Figure 18 shows a typical solar system that is designed to fulfill the hot-water needs of a four-person household. It is the most widely used type and has found application in all but the least sunny climates. The system is composed of a collection loop that amasses and transfers heat, and a delivery loop that receives the heat and delivers it to the house in the form of hot water at the desired temperature.

The collection loop consists of two 4-by-8-foot flat-plate solar collectors, an expansion tank, heat exchanger, circulating pump, heat-transfer fluid, various valves, copper pipe, and fittings. The delivery loop consists of an 80-gallon insulated solar-heated water-storage tank, a conventional water heater, various valves, copper pipe, and fittings.

When the collectors are 20 degrees hotter than the water in the solar storage tank, a differential thermostat activates the pump motor and the fluid in the collection loop begins to circulate. This fluid, an antifreeze mixture of water and non-toxic propylene glycol, is sent to the upper manifolds of the collectors, and proceeds down the transfer tubes, picking up

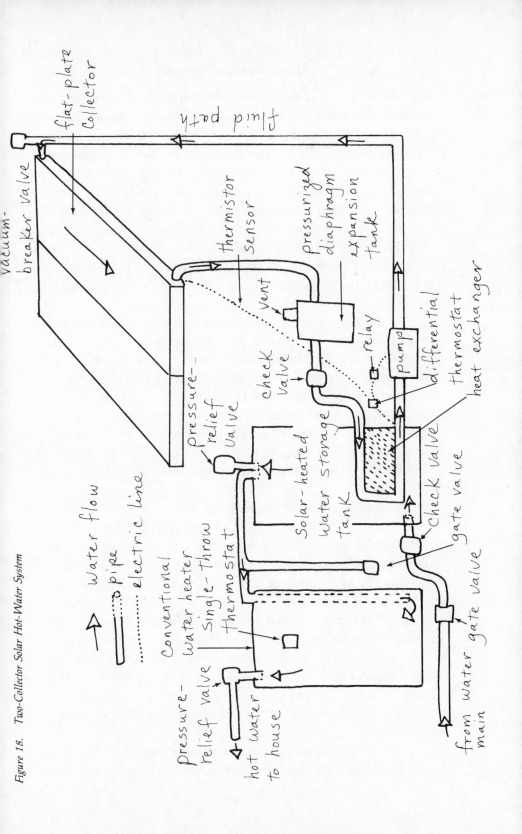

Figure 18. *Two-Collector Solar Hot-Water System*

flat-plate collector

vacuum-breaker valve

fluid path

thermistor sensor

pressurized diaphragm expansion tank

vent

relay

check valve

pump

differential thermostat

heat exchanger

Pressure-relief valve

check valve

gate valve

Solar-heated water storage tank

Conventional water heater

single-throw thermostat

Pressure-relief valve

hot water to house

from water gate valve

water main

water flow

pipe

electric line

heat from the absorber plates. After leaving the collectors through the lower manifolds, the fluid passes through an *expansion tank* that contains a synthetic-rubber diaphragm which adjusts automatically to the changes in volume and pressure caused by the rise in fluid temperature, and continues to a *liquid-to-liquid heat exchanger* where it surrenders the collected heat. The liquid is then drawn back into the pump to start a new cycle. Circulation continues until the difference in temperature between the collectors and the solar-heated water in the tank falls to 3 degrees. At this point the differential thermostat stops circulation by cutting off electricity to the pump motor via a relay.

As the fluid is heated, dissolved air is released. If allowed to accumulate in the system, this air would adversely effect circulation and hasten the destructive effects of oxidation. The system is therefore equipped with an in-line air purge.

A *check valve* in the return line immediately above the heat exchanger prevents a back flow of fluid and loss of heat. At night, for example, solar-heated water could otherwise escape by thermal siphoning to the cold collectors. The check valve permits the fluid to move in only one direction and prevents this heat loss.

Gate valves are provided in front and back of all components so that they may be isolated and serviced if necessary. Additional valves (not shown) permit draining and filling of the system. An air-vent vacuum-breaker valve is installed at the very top of the system to prevent vacuums from developing during drainage.

Both the conventional and solar tanks are equipped with *pressure-relief valves* for safety. The valves are set to depressurize the system if internal water temperature exceeds 180 degrees.

The solar tank is internally coated for rust prevention and rated for a *working pressure* of 100 psi (pounds per square inch), which is equivalent to 150 psi *test pressure* and 500 psi *bursting pressure*.

When a faucet in the house is opened, hot water is delivered at 145 degrees from the conventional water heater. The temperature is controlled by a thermostat in the usual manner. Since the entire system is under 35 psi—normal household pressure—an amount of water equivalent to that being withdrawn is sent from the solar to the conventional tank. Simultaneously, an equal amount of water is drawn into the solar tank from the city main or well.

If the solar-heated water is below 145 degrees, the temperature of the water in the conventional tank will drop and the thermostat then activates the conventional heat source to raise the water to the desired temperature. If the solar-heated water is 145 degrees or higher, no temperature drop will occur and the conventional heat source remains off.

Cold water is drawn into the solar tank and, as soon as the temperature drop results in a 20-degree difference between the water and collectors, the differential thermostat starts fluid circulation and heat transfer occurs.

A check valve between the solar and conventional tanks prevents colder water in the solar tank from mixing with hotter water in the conventional tank.

The system remains in use throughout the year and is drained only for servicing. The amount of propylene glycol in the transfer fluid is matched to the local climate, eliminating the problems of freezing and burst pipes.

Performance

During four years of operation, the system has functioned satisfactorily and continues to accomplish its purpose of providing hot water at 145 degrees, partly by solar and partly by conventional means. Thousands of similar systems are in use throughout the country with similar results.

One of the major factors contributing to the good performance of this system is the ratio of solar-absorption surface to the work it must perform: each collector has to provide heat for 40 gallons of water compared to 93.7 gallons in the space-heating system. The favorable ratio in the hot-water system leads to consistently higher stored-water temperatures and a much better utilization of the amassed heat.

Table 3

Company	System Model	Tank Size
American Solar Heat	NE 128-80	80
Groundstar Energy	2058	120
Daystar	HW3/F-BD Double	80
Revere	79543-3	80
Daystar	HW3/F-C Single	80
Sunworks	PAK 1000/4	65
Sunworks	PAK 1002/3F	120
Columbia Chase	Package 1A	65
GED	HUD 2-A	66
American Solar Heat	NE 96-120	120
Suntap	3-81	80
Intertechnology	Joule Box 3	65
Grumman	60AS3/60AST3	82
Sunworks	PAK 1001/3F	80
American Solar Heat	NE 96-80	80
Sunworks	PAK 1000/3	65
Grumman	60FS3/60FST3	82

In the summer, despite continuous heat loss due to seepage, solar hot water provides 85 percent or more of the total required. During winter, as might be expected, the solar yield drops to between 20 and 25 percent, an amount that is nevertheless a good deal higher than that for space heating. During the rest of the year, yields are between the two extremes.

In addition to the greater proportion of heat input into the hot-water system, higher utilization is due in part to the better efficiency of an immersed liquid-to-liquid heat exchanger, as opposed to a liquid-to-air heat exchanger.

Another major contributing factor is the average 55-degree temperature of water entering the system. At worst, even when the stored water drops to 70 degrees, the difference is still 15 degrees and a significant amount of heat transfer occurs. Most often the 55-degree incoming water is being heated by fluid that is 50 degrees hotter or more. The relatively large difference markedly increases heat-exchange efficiency.

Table 3 (cont'd)

Number of Collectors	Conn.	Vt.	N.H.	Mass.	R.I.	L.I. *	Del.
			Percent Solar				
4	62	55	56	55	64	61	67
3	60	50	52	53	61	56	65
3	59	48	51	52	59	55	64
3	58	47	50	51	58	53	62
3	53	43	46	47	55	50	59
4	53	44	46	47	55	50	60
3	53	43	46	47	54	49	59
3	51	41	44	45	53	47	58
3	50	40	43	44	52	46	56
3	48	39	42	43	51	46	56
3	48	39	42	43	51	45	55
3		39	42	43	50	45	54
3	48	39	42	43	50	45	55
3	48	39	41	43	50	45	54
3	46	38	40	41	49	44	54
3	45	37	39	40	47	42	52
3	44	35	37	39	46	41	51

*Long Island

Still another factor that contributes to better heat utilization is the relatively shorter time required to raise the temperature of the solar-heated water. Since only 80 gallons is stored and there is a relatively high absorption surface, significant rises in water temperature occur within a couple of hours of sunshine. Importantly, the accumulated heat is not withdrawn continuously in small amounts. Most hot water is used in the early evening, after the collection period has ended and before much heat has been lost by seepage. The result is not only better collection, but better utilization as well.

During four years of operation, solar hot water has provided an average of 60 percent of the total hot water used by the four-person household.

The Department of Energy recently conducted performance tests of various solar hot-water systems and the results are presented in Table 3 to provide a wider picture of what can be expected from other installations. ("Percent solar" numbers indicate the percentage of the average yearly hot-water requirements of a family of four that is supplied by the solar system at the indicated locations.)

Variations in performance shown by the results are mainly due to the use of aluminum absorber plates or copper ones that were too thin. Since the tests, manufacturers have switched to thicker copper absorber plates (.009 inch) and have thereby raised efficiency an average of 11 percent. The homeowner can therefore expect better performances from even these less efficient systems and a yield that is close to the 60 percent of the system under discussion.

Cost-Effectiveness

A four-person household uses an average of 75 gallons of hot water per day at an approximate yearly cost of $350 (at 60¢ per gallon). If 60 percent is solar-supplied, fuel savings at current prices amount to $210 yearly. Part of the savings is offset by a $30 yearly operating and maintenance cost, leaving a net balance of $180.

The initial cost of the system was $2,000. A twenty-year life-span is projected. Allowing for a 10 percent yearly rise in fuel costs, we have the following economic picture:

Table 4

Years	Annual Fuel Savings	Cumulative Total
1	$180	$180
2	198	378
4	240	836
6	290	1,390
8	351	2,060
10	425	2,871
12	515	3,854
16	755	6,486
18	914	8,231
20	1,106	10,342

Table 4 shows that by the time longevity has expired, the cumulative total of fuel saved amounts to $10,342. During this same twenty-year period, had $2,000 been placed in a savings account, interest and capital would have risen to a total of $10,850. The system has therefore failed to amortize itself by $508.

This is a small amount over a twenty-year span, particularly since our forecast assumes a complete loss of the system. More likely it will continue to perform longer, and one additional year will swing the balance in favor of the solar system.

There is also a realistic possibility of installing the system for less than $2,000 by comparison shopping and/or owner labor (without introducing inferior components). An initial savings of $200 will make the solar system economically viable today.

Another advantage for solar hot water (as well as other solar-energy installations) is the tax savings: a federal tax credit, and rebates in many states. Check with your tax accountant on such savings available to you.

In coastal locations such as Long Island, Rhode Island, and Connecticut, and in more southerly areas, yields are higher than 60 percent; even at an initial cost of $2,000, solar hot water is cost-effective compared to the conventional water heater. In more northerly and inland locations where yields are generally less, solar hot water is cost-effective if an electric hot-water heater is used or the price of fuel oil is more than 65¢ a gallon.

Today, improvements in the efficiency of solar hot-water systems are vigorous and ongoing: systems that supply more than 60 percent of the hot-water needs are likely to become the rule rather than the exception. Solar hot water, already cost-effective in most regions of the country, is almost certain to become more so with wider applications in all areas except those where there just isn't enough sunshine to make it work. The homeowner should not expect a significant financial reward and only small savings will be realized over the long run. However, strong environmental and energy considerations make the move toward solar hot water highly desirable.

Solar Hot Water for Pool and Household

Swimming pool water is heated both for comfort and to extend the swimming season. Conventional gas heaters are usually employed, and the cost of heating thousands of gallons of water is less than one might think because the water temperature has to be raised only a few degrees. Therefore, an average cost of $40 per month over the swimming season is not unusual. Alternatives such as solar flat-plate collector systems, reflective panels, and greenhouse enclosures have been developed, but many of these perform unsatisfactorily or are not cost-effective. Quite often an already-present gas heater performs well and the homeowner cannot economically justify a change.

Figure 19 shows an owner-built *solar hot-water system* that was installed after gas had been used for many years to heat a 16-by-40-foot, 20,000-gallon swimming pool. The system also supplies hot water to the house throughout the year. The initial cost is a great deal lower than the cheapest alternatives and the installation has a projected longevity of thirty to fifty years under continuous outdoor use. Components are available through lumber yards or at solar-equipment dealers and were assembled in a weekend.

The main component is an extrusion of a flexible black synthetic rubber, *EPDM* (ethylene-propylene-diene-monomer), which both collects and transfers the solar heat. The material is extruded into continuous lengths 4 inches wide, containing six *transfer tubes* joined to flat absorptive surfaces. (The transfer tubes are exposed to the sun and also provide direct absorptive surfaces.) Twenty parallel 100-foot-long strips provide a total of 120 transfer tubes. The installation is 8 by 100 feet and has a total absorptive surface of a bit less than 800 square feet. The transfer tubes are connected to manifolds at each end of the system. The manifold at the left admits cold water and the one at the right receives heated water. Tube ends are connected at predrilled manifold holes by a sleeve-and-grommet fitting and simple insertion of the tube end.

The collector cover is constructed from twenty-five fiberglass sheets, each 4 feet by 8 feet by $\frac{3}{16}$ inch. Wooden furring strips measuring 1 inch by 3 inches are attached with mastic and screws to the 4-foot sides of each sheet and to the 8-foot sides at each end. The 8-foot edges of adjacent sheets along the array are contained in 8-foot-long aluminum channel strips with grooved sides and are sealed with neoprene.

An insulated back is provided by 2-inch-thick Styrofoam sheets. EPDM strips are attached to the surface of the Styrofoam with mastic.

The installation is located several yards from the pool and faces true south. Stakes and earth along the upper side support the collectors at a 32-degree angle, which is 10 degrees less than the latitude of the site and is the optimum angle for summer use. Additional support is provided by a parallel pair of intermittently staked 2-by-4s that run the entire length under the Styrofoam sheets.

Cold pool water is supplied from a tee beyond the filter in the pump-outlet pipe. A *flow valve* immediately past the tee regulates the flow of water to the collectors. (Fittings and pipe are plastic.) The supply pipe is joined to the left manifold. As water rises in the manifold and circulates in the transfer tubes,

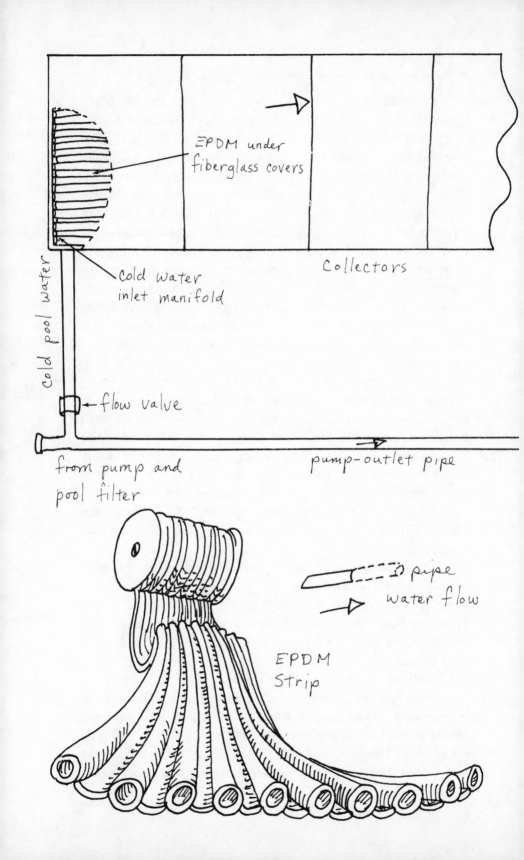

EPDM under
fiberglass covers

Collectors

Cold water
inlet manifold

cold pool water

flow valve

from pump and
pool filter

pump-outlet pipe

pipe
water flow

EPDM
Strip

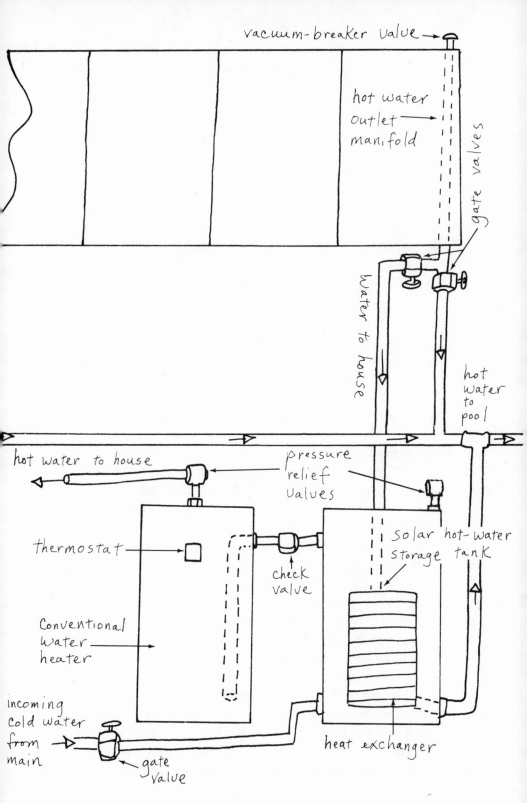

Figure 19.   Solar Hot-Water System for Pool and Household

solar heat amassed by the tubes and flat areas is transferred, and heated water spills into the right manifold. This heated water leaves through a connecting pipe, rejoins the pump-outlet pipe, and empties into the pool.

Several other valves permit diversion of heated water to the house. The hot water is run through a liquid-to-liquid heat exchanger inside a 150-gallon solar storage tank and transfers its heat to the household water. The solar tank is connected to a conventional water heater and the water is delivered in exactly the same way as described in the chapter "Solar Hot-Water System" on page 36.

In above-freezing weather, after leaving the heat exchanger the pool water is piped to the inlet manifold of the collectors and reused. In freezing weather, after the pool is drained, water for the collectors is supplied by two 55-gallon drums adjacent to the solar storage tank located in the garage. At all times the collector loop is unpressurized, while the household water is pressurized at the normal 35 psi.

Performance

EPDM is only a fair absorber and conductor of heat, and approximately 20 percent less efficient than copper as a heat-transfer material. The initial loss is offset by a number of positive factors. In the other solar systems previously described, hot water from the collectors undergoes two heat transfers with attending losses. The losses are particularly high in the space-heating system, since hot water is stored for a longer time. In the pool system there is only one heat transfer, and the heated water is used immediately. Therefore, despite an initially poorer heat transfer (from sunshine through EPDM), overall efficiency is a good deal higher than in other common solar systems.

Efficiency is also improved because the pool water is heated when air temperatures and sunshine are at their highest. The 800-square-foot absorber surface is quite large for the intended function and also contributes to good performance. Sunshine directly striking 640 square feet of pool water surface also makes a substantial contribution. The combination of these factors produces a highly workable system that raises pool water temperature an average of two degrees per hour. (Theoretically, 166,600 btus are required to raise pool water 1 degree, but an additional 20 percent is needed to replace heat simultaneously lost by convection from the surface of the water.)

The pool begins to be used in early June. Temperatures at the start of the day are generally in the high 50s and pool water is a few degrees lower due to heat losses during the night. Losses can be minimized by covering the surface with plastic "water lilies" (thin plastic disks, usually 6 to 12 inches in diameter that float on the surface of the water and provide an insulating cover, substantially reducing the convective heat losses that would otherwise occur). Using the solar source alone, seven hours of continuous sunshine is needed to raise the water temperature from an assumed 56 degrees at the start to 70 degrees (the lowest comfortable level for most people). During this spring period and similar periods in the fall, the gas heater is used along with the solar source to speed up heating so that 70 degrees is reached by early afternoon. The gas heater is then shut off and further temperature rises are brought about by the solar source alone.

As the swimming season progresses, daytime temperatures climb into the middle and high 60s; pool water that has been in the 70s the day before and "insulated" during the night fluctuates around the ambient temperature at sunrise. If there has been an appreciable wind during the night, the pool water is generally 3 to 5 degrees cooler. In either case, pool water temperature is raised well above the minimum before noon and by four o'clock it is generally in the upper 70s. The gas heater is rarely used during this period except after rainy days, and then is used only intermittently.

During summer months the solar system alone provides all the heat necessary to maintain the pool temperature in the high 70s and low 80s. Quite often it is necessary to shut off the system to keep the pool water from becoming uncomfortably hot.

EPDM withstands continuous temperatures of 375 degrees and is not affected when the system is shut off. Although interior temperatures in the collectors reach 300 degrees at times, a safe margin remains and nothing needs to be done to cool the material.

When the system was first installed, the rate of flow was measured with the flow valve in various open positions. Water was run through the collectors and deposited in a 55-gallon drum. Filling time was noted with a stopwatch and a corresponding mark placed on the flow valve. The temperature was also taken each time, and the most effective rate of flow was determined by multiplying the temperature of the water and the rate of flow, then comparing the totals. The optimum rate of flow over the entire swimming season was found to be 10 gallons per minute (gpm). (The owner considered it too troublesome to change the setting of the valve each time the ambient temperature changed and traded off a bit more efficiency in favor of simply leaving a 10 gpm flow of pool water all the time.)

In above-freezing weather, the valve is periodically restricted to 2 gpm and hot water is routed to the household system. Water temperatures entering the heat exchanger are usually 150 degrees or higher and heat transfer is excellent. It generally takes a couple of hours to fill the solar storage tank with 145-degree water. The hot water lasts a few days, then the process is repeated.

In freezing weather, a small circulating pump in the garage draws cold water from the two 55-gallon drums and pumps it through the collectors and heat exchanger and then back to the drums. When the water temperature in the solar tank reaches 120 degrees, or higher if weather permits, the pump is shut off. Water in the transfer tubes drains from a high point at the center to the manifolds and back into the drums. An air-vent vacuum-breaker valve prevents a vacuum from developing. Some water remains in the tubes and freezes during the night, but no damage is done because EPDM has a high elongation factor of 350 percent and simply expands under the ice's pressure. (EPDM also withstands temperatures 80 degrees below zero without deterioration.) With the appearance of the sun in the morning, the ice inside the tubes melts and the collectors are again ready for use.

This multi-use system is remarkably simple to operate: for pool use, the flow valve is simply opened to the 10-gpm mark. To transfer the collectors to hot-water heating in above-freezing weather, the valve is turned to the 2-gpm mark, the valve to the pool is closed, and another one to the exchanger is opened. In freezing weather, another separate loop is employed and the small circulating pump is simply switched on.

Most often there is more hot water available than can be used; over the course of a year the solar system supplies about 75 percent of the requirements of the four-person household. The conventional water heater is primarily used to maintain the water at 120 degrees and occasionally after periods of consecutive overcast days.

Cost-Effectiveness

The total cost of the system at current prices is $2,200. Fuel savings amount to $180 yearly for the pool and $270 yearly for household hot water, a total of $450 yearly. Operating and maintenance expenses are limited to the small circulating pump and are less than $10 yearly. The pool pump is not included in the cost since it is necessary in any case for filtering and performs no additional work because the solar system is present. Net value of fuel saved is $440 per year.

Assuming a 10 percent yearly increase in fuel costs set off by an 8 percent interest return on $2,200, at the end of the fifth year the value of fuel saved is $3,381, compared to $3,232 of accumulated capital and interest. The installation is amortized; during the remaining twenty-five to forty-five years of its projected life-span it continues to provide hot water for the pool and household at fuel savings of 90 percent and 75 percent respectively.

A relatively low initial outlay due mainly to less expensive EPDM is an attractive feature in switching from the exclusive use of gas heaters. The high degree of cost-effectiveness combined with the simplicity of installation and operation makes this system desirable in virtually all climates. It isn't often that a home improvement does everything expected of it while saving a considerable amount of money. This system is undoubtedly a most welcome addition to the energy-saving scene.

Passive Solar Heat/The Insulated Shutter

Installations that use solar energy without recourse to mechanical equipment (except fans) comprise the *passive solar* category. The ordinary window is the most ubiquitous example, and when facing south it is far more efficient than a collector in utilizing solar heat. During the heating season in sunshine-poor regions, windows admit an amount of solar heat that is equivalent to 10 percent of the heating bill. In regions such as the Sierra foothills, where winter temperatures rise into the 70s by day and drop to the 30s at night, the equivalent can be as high as 50 percent. The amount of heat involved in all climates is significant, so its utilization is a major energy-saving factor.

Windows, however, are also the major cause of heat loss in a house and often account for 60 percent or more of the heating bill. Heat is lost through them at a rate five times greater than for adjacent insulated walls, and under windy conditions, even greater loss occurs.

Passive solar installations must increase the amount of solar heat admitted and decrease heat losses without affecting the input. Solar input can be improved through the use of glass with high transmittance and low reflectance, but the gain is minimal. Furthermore, unless we are dealing with a house still to be constructed, the cost of replacing the existing glass is too high for the value returned. So, for all practical purposes, improving the window as a passive solar heat source means decreasing heat losses without affecting input.

Storm windows have been one of the traditional means of cutting heat losses and do so by approximately 40 percent. Welded glass, constructed from two panes with a partially evacuated air space between them, is also used extensively and cuts heat losses by about 35 percent. Thermopane—two sheets of separated glass without the partial vacuum—reduces heat losses slightly less than welded glass. Triple glazing, which involves three panes separated by air spaces (now just beginning to be used here but long a standard installation in European Alpine regions), is most effective and cuts heat losses by 57 percent. However, all these improvements reduce the amount of solar heat admitted to the house, and leave major heat losses. They are also expensive—in the case of triple glazing, prohibitively so at the present time.

The use of multiple glass represents an advance that was introduced by manufacturers and suited their needs rather than those of the homeowner. As a result, these windows cost more than the heating system or siding and have become the single most expensive item in a house. The *insulated shutter* shown in Figure 20 is a far less costly approach. Much more efficient in heat retention, this device is well suited to the needs of the homeowner.

The shutter is composed of four 1-by-3-inch wood strips for the frame and a front and back of $\frac{1}{4}$-inch tempered Masonite. After assembly, a hole is bored in one of the wood strips, the cavity is filled with polyurethane foam insulation, and the hole is plugged with a dowel. The shutter is mounted on the interior side of the window jambs with standard hinges. T-shaped plastic weatherstripping around the perimeter is used to ensure a tight fit and prevent infiltration losses.

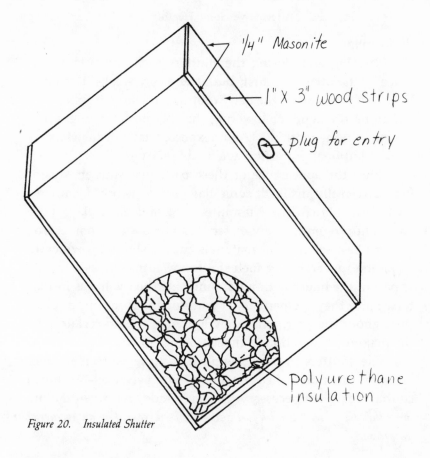

Figure 20. Insulated Shutter

Making the shutter takes less than an hour for a person with minimum carpentry skills, and mounting it is easier than hanging a door. All materials are available at any lumberyard for a very modest cost.

When the shutter is closed, the heat loss from the glass area becomes less than that of an equivalent area of adjacent insulated wall. Thus, *the major source of heat loss in a home is eliminated.*

The shutter is best employed in conjunction with a single pane because ordinary windows admit more solar heat than do multiple panes. The substantial cost of multiple-pane installation is avoided, while *the degree of heat retention is far higher than with triple glazing.*

Performance

Opening and closing the shutters each day was a nuisance to the owners at first, but after a few weeks it came to be regarded as part of the daily routine—no more trouble than switching off lights or lowering the thermostat at night and raising it in the morning. (An unexpected asset of the shutters is the additional protection against burglary.)

Since the appearance of these particular shutters was at first universally disliked, some have subsequently been covered with fabric. (The Masonite front and back plays little part in heat retention; any material can be used.) A homeowner has a vast selection of materials and finishes with which to make the appearance of such shutters as attractive as any other part of the house without diminishing their value as insulators. Since they are permanent additions that remain in place throughout the year, what they look like should certainly be the major consideration in their design.

The shutters in our example case have performed their essential function of heat retention extremely well—so much so that it is often necessary to open windows for ventilation. *Since they have been installed fuel costs have been reduced by an average of 40 percent.*

Cost-Effectiveness

Materials and hardware for the shutters purchased at retail prices averaged less than $15 per window. (A local mill was willing to make the shutters and supply hinges for $17.) A carpenter offered to mount the shutters at an average of $30 each. (The owner chose to do the work himself.) A total average cost of $47 per shutter for materials and labor would be a reasonable estimate.

The house has twenty windows and the cost of having them shuttered was $914. It also has an oil furnace; the heating bill for the year preceding the shutter installation was $937. In the first year after the shutters were installed, the bill dropped to $548. During the same year, the price of oil rose, so that the total without the shutters would have been $1,021. In the first year of operation, the shutters saved a net of $473. (Of course, there is no operating expense.) In the second year, the savings in fuel completely amortized the initial outlay; since that time the shutters continue to cut fuel bills by more than 40 percent.

There is no single type of home-heating improvement that even approaches the cost-effectiveness of insulated shutters. There are no moving parts to maintain or wear out, no fuel expended, no complicated controls to malfunction. Shutters are easily fabricated from universally accessible materials, are far less expensive than multiple-glass alternatives, and are much more efficient. They can be installed one at a time if finances so dictate, and the considerable benefits are permanent. For the above reasons, insulated shutters are wholeheartedly recommended to the homeowner.

Passive Solar Concrete Wall

The fact that masonry materials absorb and emit heat slowly has long been known to occupants of New England stone houses. The installation of a masonry wall to provide space heating through solar radiation was proposed in the nineteenth century. The wall was conceived to serve the structure as any wall does, hence most of the expense in building it was assigned to ordinary construction costs. The sun was the free heat source; there were no moving parts, no operating expenses, and no work was required of the occupants. The installation was permanent and benefits continued indefinitely.

The idea did not find practical expression until the 1960s when solar walls began to be built in England and France, as well as in our own Southwest and—more recently—in the Northeast and climatically similar regions. But there has been no stampede to build solar walls; their total today numbers no more than several hundred.

Figure 21 shows a modern *solar wall installation*, a widely publicized type that is often synonymous with passive solar use. The system is located in the Northeast and is designed to fulfill some of the space-heating needs of a two-story 1,800-

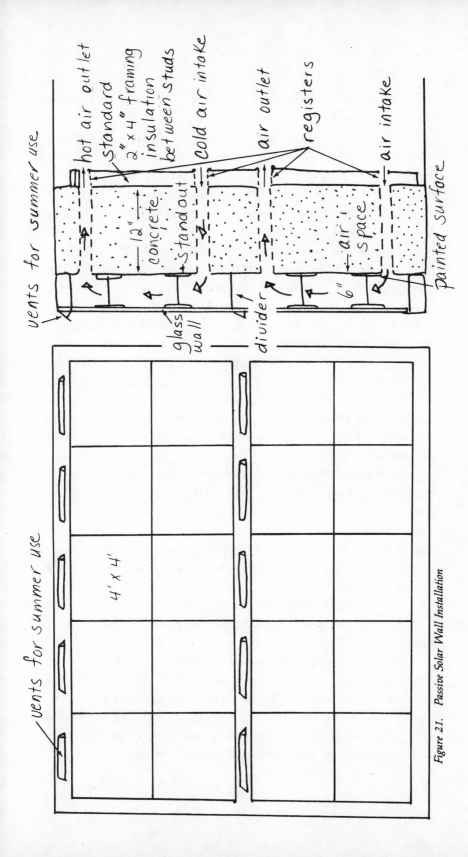

Figure 21. Passive Solar Wall Installation

square-foot house. The owner acted as contractor and all labor and materials were obtained locally.

The masonry component is a concrete wall, 1 foot thick, 20 feet long, and 16 feet high. The front of the wall, the *solar absorptive surface*, is painted black and faces true south. The back of the wall faces the living room on the first floor and two bedrooms on the second; it is insulated and finished in the same way as adjacent walls.

Outside the wall, *double-strength glass panes* 4 feet square are mounted in an aluminum framework of standard glazing channel. The framework is secured to the concrete wall by *standouts* (bracketlike hardware), and the two are separated by a 6-inch air space. Aluminum strips around the perimeter make the assembly airtight.

The air space is divided into upper and lower halves by an additional horizontal strip in a plane with the second-story subfloor. Directly below the dividing barrier are five vents spaced along the 20-foot length; five other vents serve the top half of the wall. These vents open outward to evacuate hot air in summer.

Four adjustable registers are spaced along the living-room wall a few inches above the floor. Openings behind the registers continue through the wall and permit room air to be drawn into the 6-inch air space. A few inches below the ceiling, four additional registers permit heated air to leave the air space and return to the living room. A similar arrangement serves the second floor.

A variable-speed fan near an upper, central register in the living room (and each of the bedrooms) provides forced circulation.

On a sunny day about half of the short-wave radiation available passes through the glass and is absorbed by the painted concrete surface. Unlike a selective coating, the painted surface has a high degree of emissivity, so that most of the absorbed heat is returned to the air space. A small portion is conducted toward the interior of the concrete wall.

About 15 percent of the radiation that strikes the painted surface is reflected as longer waves; these accumulate in the air space and give rise to the superheated greenhouse effect.

As cool air strikes the painted surface of the wall and the inner surface of the glass, it absorbs heat and begins to rise. Air from below rises to fill the space that is left, and natural circulation begins. The hottest air exits through the upper registers and relatively cool air from the room is drawn through the lower registers into the air space.

The movement of air in the 6-inch space is not a straight line from bottom to top. Although the general direction is upward, each portion of air that strikes a heated surface is also deflected laterally and creates turbulence. Variations in temperature over the inner glass surface, concrete, and different portions of air also contribute to turbulence. The 6-inch space is much wider than spaces in solar collectors, so turbulence is a good deal stronger.

Greater turbulence has the advantage of increasing the frequency of air contact with heated surfaces and therefore results in better heat transfer. However, turbulence also causes substantial conductive heat losses through the glass as well as convective losses around the perimeter. On windless days the heat loss is about 20 percent; with a wind of 30 mph or so, twice as much is lost.

Apart from its solar function, the concrete wall encloses a portion of the house and supports the second story and roof. When the solar system isn't in use, the insulated windowless concrete wall provides a better thermal barrier than standard insulated walls.

Performance

During winter nights the system loses its heat except for a small amount seeping into it from indoors; at sunrise its temperature is approximately the same as the ambient one. Virtually all radiation from the eastern sun strikes the glass tangentially and bounces off. As the radiation angle improves, the glass, the painted surface, and the air between them begin to warm to operating level. During this morning period the system does not provide heat to the house and registers remain closed.

By midmorning, after the position of the sun reaches 150 degrees (true south is 180 degrees plus or minus magnetic declination), effective absorption of radiation begins. Within a short time, usually before eleven o'clock or so, temperatures in the air space have risen to between 85 and 90 degrees, and registers are opened. The heat transfer has occurred, and from this point until late afternoon on an average winter day, there is generally enough heat to keep the 20-by-26-foot living room and the bedrooms at a 70-degree temperature. On a windless day with the temperature in the upper 30s, a small excess of heat is produced. If the temperature is below freezing and the wind 15 mph or so, the amount of heat produced is insufficient to maintain the 70-degree temperature in the living room and bedrooms, and the furnace must be used. Over the course of a heating season these factors tend to cancel one another out and the system produces enough heat to maintain the living room and bedrooms at 70 degrees during sunny days.

As the sun begins to go down, the system cools quickly. In less than an hour the concrete mass has emitted all its effective heat, and the registers are closed until midmorning of the next sunny day.

In early and late periods of the heating season, there are many afternoons when an excess of heat is produced and fans are used to distribute it to other spaces in the house. (The cost of fan operation is much less than the value of the heat.) Under optimum conditions of continuous sunlight and ambient temperatures in the low to middle 60s, the solar wall provides enough heat to maintain about half the house at a 70-degree temperature.

Throughout sunny periods the buildup of heat within the concrete wall is slow, and insulation confines most of the heat to the air-space side. Since heat transfer occurs within the air space, both the insulation and high emissivity are desirable, and the concrete mass transfers little heat directly to the house.

Cost-Effectiveness

The 320-square-foot concrete wall cost $2,450. Since the wall also serves as an exterior structure, its value as such is deducted. The other exterior walls of the house average $5 per square foot; if the same cost is applied, $1,600 is attributable to normal construction expense and $850 to solar use.

The system saves a net of $70 yearly in fuel costs. If we assume a 10 percent yearly rise in the price of fuel and an 8 percent interest return on $850 in capital, the net yearly saving is $9. Obviously, amortization is a very long way down the road; however, the system is permanent and therefore cost-effective, even if only barely so.

A permanent installation that costs nothing to use and requires only the opening and closing of registers and occasionally switching on a fan is a positive addition to a home. The system would be much more desirable if its simplicity could be retained with cost-effectiveness substantially increased.

In examining the system with a view toward improvement, one of the first negative features to become apparent is the minor role of the concrete mass. Since only its surface plays an important part in heat transfer, other surfaces, without concrete's mass and attendant costs, could easily be used instead.

Another negative characteristic is the substantial heat loss caused by the relatively wide 6-inch air space.

Apart from efficiency or economic considerations, the massive wall creates a large lightless area inside the house. The wall converts a heating mechanism into a major determinant of the house's appearance, and many homeowners will find this aesthetic factor unacceptable.

Solar researchers and equipment manufacturers have addressed themselves to these problems and have developed alternatives. One such installation that presents problems of its own employs ordinary-sized passive solar collectors as wall modules. A fiberglass cover and painted metal absorption plate establish a 2½-inch air space in which solar heat is absorbed and transferred to interior air. Insulating Styrofoam pellets are pumped in and out of the air space as needed. This system, and others similar to it, share the same basic deficiency of transferring heat in the same area that collects it, and at the same time cancel out the advantages of the greenhouse effect by high turbulence and attendant heat losses. Pumping pellets complicates the system needlessly and reduces cost-effectiveness below the break-even point. In all of these systems, the air space cannot be reduced without also reducing the circulating air to an ineffective amount.

Figure 22 shows an experimental passive *solar collector* that is designed to resolve the problems connected with the concrete wall and also avoid the deficiencies in other passive systems.

The collector is 4 by 8 feet and has a flange around the edge for direct nailing into the studs of a conventionally framed wall. The cover is a 4-by-8-foot sheet of translucent fiberglass, and the absorber plate, attached 1 inch behind it, is sheet copper with a selective coating. The back side of the absorber plate forms one side of a rectangular metal duct, 4 by 8 feet by 3 inches deep. Behind the duct, standard 3½-inch fiberglass insulation fills the areas between studs in the normal fashion and eliminates the need of pumping loose insulation. At the bottom and top of the duct are pairs of tees capped by registers at the finished exterior and interior walls. (The exterior registers are left open in summer to evacuate hot air.)

The 1-inch space between the cover and plate substantially reduces turbulence and attendant heat losses. The selec-

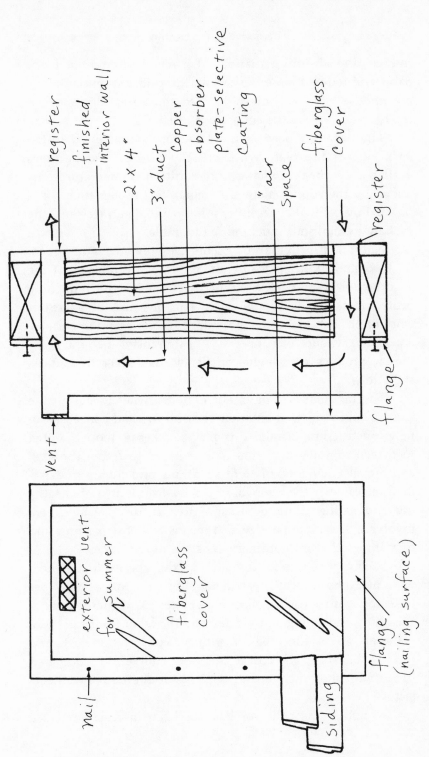

Figure 22. Modular Solar Wall Collector

tive coating absorbs radiation 10 percent better than black paint and reduces emissivity to an insignificant amount. The greenhouse effect is entirely retained and a great deal more of the heat that enters the collector is utilized.

The heat produced in the collector is transferred to the back side of the plate with virtually no loss and becomes available for heat exchange. In addition, temperatures attained by the copper plate are a good deal higher than those reached by painted concrete or other painted surfaces, making exchange conditions much more favorable.

Natural air circulation occurs as previously described; heat exchange takes place by air contact with the back of the absorber plate and, to a minor extent, with the remaining heated metal walls of the duct. The presence of a greater amount of heat, higher temperatures, briefer warmup periods, and the excellence of copper as a heat-transfer material combine to produce a far higher efficiency than concrete-wall installations.

The modular size permits the homeowner to design the south wall in any combination of windows and collectors. In new construction, installing a collector takes no more than ten to fifteen minutes, less time than needed to frame the wall and install a window, or to fit sheathing and siding. In existing houses, only the removal of a 4-by-8-foot area of sheathing and siding is required, and no structural changes are involved. The modular size also means less cost per unit and easy integration into ordinary construction practices.

Half a dozen collectors such as the one evaluated here were made by a local sheet-metal shop at a cost of $147 per collector. Since each collector replaces 32 square feet of sheathing and siding and virtually all the labor in fitting these materials, $64 is allocated for ordinary construction costs and $83 for solar use. The cost is $2.59 per square foot compared to $2.65 for the concrete wall or $3.90 for alternative wall collectors.

Although the collector has not been in residential use

long enough to provide reliable performance data, the soundness of the design and performance under simulated conditions points very strongly to the conclusion that it is far more cost-effective in utilizing passive solar heat than the concrete wall and other alternatives. The most conservative estimate is that each collector will amortize itself by fuel saved within ten years; the most optimistic estimate reduces that time by half.

The wall collector retains the decided advantages of simplicity and permanence common to passive systems. It costs less and is easier to install than a glassed-in concrete-wall system and is almost certainly a good deal more efficient and cost-effective. The unit can be easily made from standard materials by any sheet-metal shop or solar-equipment manufacturer. The modular solar wall collector has been presented here with the conviction that it is the best way to use passive solar heat today and deserves to come into widespread use.

Solar Greenhouse

The atypical *greenhouse* shown in Figure 23 is designed not only to grow plants throughout the year but also to serve as a living space and source of passive solar heat.

The floor is a concrete slab covered by quarry tile—masonry materials that absorb and release heat slowly. The framework is aluminum and the glass single-strength with a high solar-transmittance value. Three 2-foot-thick stone-and-mortar walls rise to a height of 3 feet on the south, west, and east sides. The north wall, shared with the house, is also of 2-foot-thick stone and mortar and extends from floor to roof. The wall is painted chocolate-brown for better heat absorption. It adjoins the kitchen and living room along its rear, and two pairs of glass doors near each end provide access. Another glass door in the west wall leads to the exterior-Five spaced glass panes high on the south wall open outward to let out hot air in summer. Registers are installed at the top and bottom of the north wall through to the kitchen and living room. An exhaust fan and ducts recessed in the kitchen ceiling are joined to conventional ducts that serve the entire house.

A large movable ceiling cover is provided. The cover is

made of long, narrow insulated panels hinged together in the same manner as standard 16-foot garage doors and operates electrically with identical hardware. When the insulated cover is fully retracted, it lies in a hollow area above the second story subfloor and is hidden by a raised finished floor. When fully extended, the cover lies several inches below the glass roof and behind the south-facing wall. Hinged insulated panels serve the side walls.

An additional exhaust fan, not included in the original construction, has been installed in place of a pane in the south wall for summer use.

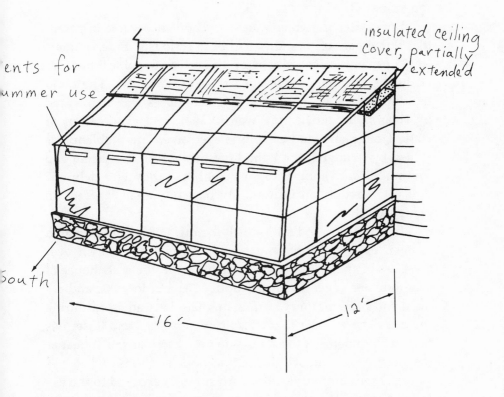

Figure 23. Solar Greenhouse

Performance

The greenhouse fulfills expectations as an area for growing plants. As a living space, it is regularly used for meals and entertaining friends.

At the start of the heating season, temperatures in the greenhouse rise well above 80 degrees on a sunny day. Ambient temperatures are generally in the high 50s or low 60s, and there is relatively little demand for space heating. Natural circulation through open registers in the north wall transfers some excess heat from the greenhouse, but the exhaust fan is needed to keep the greenhouse from becoming uncomfortably hot. During this period the greenhouse provides enough heat to maintain the kitchen and living room at 68 degrees, and sometimes other areas as well.

Soon after the greenhouse was built it became apparent that the impact of humidity had been seriously underestimated (35 percent is generally accepted as "comfortable," but levels in the greenhouse exceeded this considerably). The rapid buildup of heat characteristic of greenhouses was augmented by solar radiation heating water vapor, and evaporation of water from plants and earth was excessive. The problem was solved by removing the humidifier in the furnace system and routing the air coming into the greenhouse through a dehumidifier. The reduction in humidity had only a small effect on plant growth.

In the spring and fall, shortly before sunset, the insulated cover is extended fully and insulated panels are drawn over the glass of the side walls. Temperatures in the greenhouse at this time are generally in the upper 70s and low 80s and remain so as the masonry surfaces radiate heat amassed during the day. The exterior low masonry walls—and to a lesser extent the perimeter of the floor—do not absorb as much heat as the north wall, and within half an hour or so lose what was gained; therefore, these areas with surfaces exposed to the exterior become the major source of heat loss. Heat from the north wall continues to radiate for an hour or so longer, but the amount is smaller than the heat loss and the greenhouse temperature begins to drop slowly. Throughout the night a minor amount of heat is conducted through the north wall from the kitchen and living room but is only significant when the greenhouse temperature is about 10 degrees higher than the ambient temperature. (The low walls have since been insulated and covered with tile, and the greenhouse now retains heat as effectively as an insulated windowless room.)

On a clear day in winter, a temperature of 70 degrees or higher is rapidly achieved and a high level maintained until an hour or so before sunset. During this time no supplemental heat is necessary. On partially overcast days, when temperatures drop into the 40s or 30s, the cover is extended. In the covered period, the small amount of heat from the north wall is usually enough to keep greenhouse temperatures above freezing if there is no wind. If the day is overcast or particularly windy and cold, the cover is extended and the doors to the house are left partially open to admit heat.

Over the entire heating season, the greenhouse supplies more heat to the house than it uses from the furnace. The difference accounts for approximately 10 percent of the house's total heating requirement.

At the start of the nonheating season, natural and intermittent forced-air circulation is sufficient to keep greenhouse temperatures at a comfortable level. With rising outside temperatures, a combination of intermittent forced-air circulation and partial extension of the roof cover is necessary. At the height of the summer, forced-air circulation by both fans is needed practically all the time and the roof cover is extended most of the afternoon. (During this period, a layer of superheated air is created in the space between the glass roof and cover, and the greenhouse fan runs continuously to evacuate it.) When outside temperatures reach into the 90s, both fans run continuously, the roof cover is extended to cover a good part of the south wall, side panels are closed when necessary, and the greenhouse remains shaded for most of the day.

Throughout the nonheating season, except for the hottest days, the greenhouse is as comfortable as any other non-air-conditioned space.

Cost-Effectiveness

We have noted previously that south-facing glass is an excellent source of passive solar heat provided it is also insulated to prevent more heat from being lost at night than is gained during the day. In this respect, the greenhouse is simply a large expanse of glass that admits and traps solar heat and, since it is also effectively insulated, has a solar gain that cuts yearly fuel bills by 10 percent. The net dollar value of this gain was $55 at the time the greenhouse was built. Obviously, the amount is too small to amortize construction costs. But the greenhouse is not solely a collector of solar heat, and fuel savings are only a part of the economic picture; the greenhouse also provides living space, and its cost-effectiveness can only be compared to other living spaces of the same size.

The greenhouse has a floor area of 192 square feet and cost $8,450, $44 per square foot. An additional bedroom built at the same time, for example, would have cost $7,296, $38 per square foot. Cost-effectiveness, then, depends on whether an additional bedroom does more for the homeowner than a greenhouse that is initially $1,154 more expensive and cuts fuel bills by 10 percent. (There is no way of placing dollar values on the noncommercial growing of plants and flowers.) Any space that provides a house with more heat than it consumes makes a positive contribution to energy-saving, so a greenhouse can only be regarded favorably in this respect.

The owners have found it to be a pleasurable part of their day-to-day life and a highly cost-effective living space, particularly when they compare it to their rarely used formal dining room. Whether it would also be cost-effective in relation to other living spaces in other situations can only be answered by each individual homeowner.

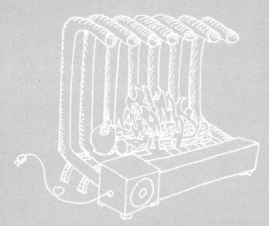

Wood-Burning Systems

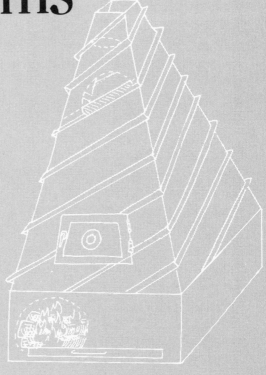

Fireplaces

The use of a fireplace as the principal means to heat a home became obsolete long before the turn of the century, but fireplaces continued to be built in large numbers, and today half the new homes have one or more. Until recently, the main purpose of a fireplace was to provide the look and feel of an open fire, with heating considerations incidental, but the rising costs of home fuels coupled with a new energy-saving public awareness have prompted homeowners to consider how more fireplace heat can be delivered to the house rather than lost up the flue, a loss that usually is close to 90 percent.

The amount of heat lost in a fireplace depends, of course, on how much the fire is producing. One pound of wood, no matter what species, yields 7,000 to 8,000 btus. Since wood is sold by the cord (a cord is a stack of wood 4 by 4 by 8 feet), not by weight, denser and heavier hardwoods are preferable to lighter softwoods. Hardwoods (the wood from deciduous trees) produce an average of 28,000,000 btus per cord compared to 22,000,000 for softwoods (the wood from evergreens). The amount of heat produced by an average fire during an hour is approximately 20,000 btus, 18,000 of which rise up the flue without warming anyone. In the course of burning a cord of hardwood in an unimproved fireplace, 1,400,000 btus are utilized at a current cost of $60. We can get the same amount of heat for $24 of electricity and only $7 of oil. Obviously it is economically unsound to burn wood in today's fireplace *unless* a significantly greater amount of heat can be utilized. Increasing the efficiency of the fireplace is not merely an academic consideration, for the 20,000 btus per hour produced by the average fire is far more than is necessary to heat a 1,500-square-foot house on all but the coldest winter days. In addition to the near certainty of rising oil prices, there are maintenance and depreciation costs of furnace use, whereas there are none attached to fireplace use. With tens of millions of fireplaces already in existence, and a great many more being built, a large-scale increase in fireplace heat recovery could have an enormous impact on the energy scene.

The last significant changes in fireplace design, splaying the side walls and slanting the rear wall of the firebox, came in colonial times, but while less smoke was blown into the room as a result, little was accomplished to increase heating efficiency. Figure 24 shows the *modern fireplace*, essentially the same as those built two hundred years ago. Still, despite the lack of change in basic design, devices were invented in the last century that succeeded in capturing a good deal of the lost heat, and some of these have provided the basis for masons, do-it-yourself homeowners, and corporate engineers to develop ways of increasing fireplace efficiency.

There are no miraculous improvements to make a fireplace economically viable in one stroke. Such viability comes only by a number of improvements, each small in itself yet decisive in combination. (All improvements are made on the premise that the open fire will be retained.)

One of the earliest innovations employs two pairs of *openings* in the chimney walls; one pair is located near the floor in the sides and the other a few feet above the firebox, either in the face or chimney sides. Cold room air lying at floor level is drawn into the lower openings, and after being heated by the throat and *smoke chamber*, it returns to the room through the upper openings. Air circulation occurs naturally and an addi-

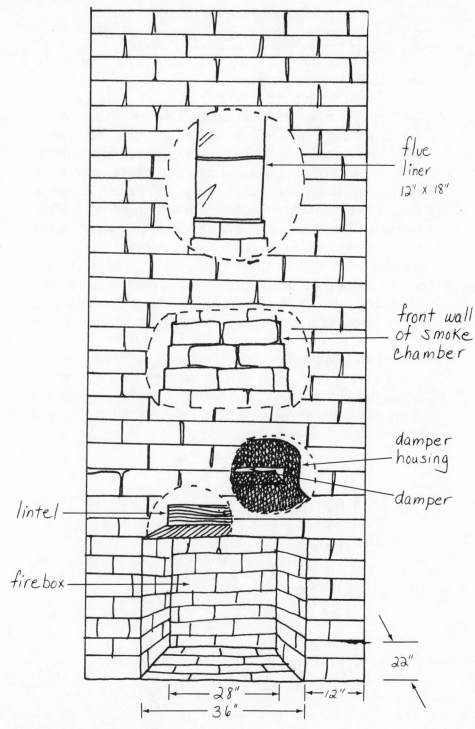

flue
liner
12" × 18"

front wall
of smoke
chamber

damper
housing

damper

lintel

firebox

28"

12"

36"

22"

Figure 24. Modern Fireplace

tional 5 percent of the heat produced is recovered. Although this is a modest amount, it costs nothing to obtain except for an insignificant initial outlay.

A metal version of the original masonry model has been marketed for many years under the Heatilator name. Air is drawn naturally into hollow chambers above the firebox, heated by gases being exhausted, and returned to the room. The use of metal results in an increased efficiency of 8 to 10 percent, but it does have disadvantages. The metal firebox is subject to the pitting action of an open fire and has a life expectancy of only fifteen years or so. The unit cannot be removed without dismantling the most labor-intensive area of the fireplace, and replacement costs are high. The unit is also a good deal more expensive initially than the masonry equivalent. So, although the use of metal is highly desirable in many locations, the firebrick alternative is superior in its cheapness, its permanence, and the fact that the additional cost of the Heatilator can be applied more advantageously.

Figure 25 shows a *C-shaped heat exchanger*, a current version of an old design. The lower sections of hollow steel tubes serve as a grate, on which a wood fire is started in the usual

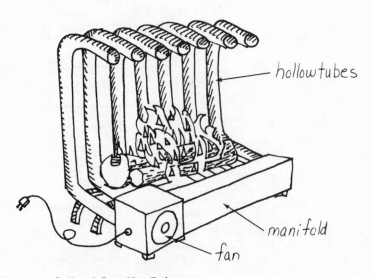

Figure 25. C-Shaped Grate Heat Exchanger

way. A small *squirrel-cage fan* at the side sucks in colder air at floor level and blows it through a manifold into the tubes, where the air is heated by the coals, flames, and hot gases rising from the burning wood. The heated air is then blown into the room through the open ends at the tops of the tubes. The ends of the tubes extend several inches beyond the front of the firebox. The pictured device was built locally six years ago at a cost of $200 and has succeeded in raising fireplace efficiency 15 to 20 percent.

The device leaves room for a good deal of improvement and can be regarded only as a first step in the right direction. Still, the amount of heat actually delivered is surprising. During a recent winter when the temperature outdoors was below freezing and a stiff north wind was blowing, a slowly burning fire on this exchanger heated a 16-by-60-foot room, and less than an hour after the fire was started, windows had to be opened because the room was too hot.

Similar devices have come on the market in the past few years at prices that range from $19 to several hundred. The cheaper versions use a grade of steel that is too light to resist pitting effectively, are not equipped with a fan, and increase efficiency by at most 5 percent. Better types all employ fans and use heavier steel. Such mass-produced fans are invariably poor in quality and improperly sized, and the exchangers are not as efficient as the one pictured. Their upper tubes are also a bit too short, and smoke is occasionally pulled into the room through the suction effect of the forced-air circulation. However, as long as wood is being burned in the fireplace, almost any grate heat exchanger will recover a significant amount of heat that would otherwise be lost, enough to pay for itself in a winter or two of use. Units are available at lumberyards, hardware stores, and national chains, and are manufactured in sizes to fit all but the most extremely sized fireboxes. There are no installation costs, only a few moments are needed for assembly, and the device retains the open fire so many of us find desirable.

The C-shaped heat exchanger, although a practical device, is a temporary expedient, since 80 to 85 percent of the heat produced is still being lost up the flue. Efficiency has to be raised to about 35 percent before the fireplace becomes competitive with the oil furnace.

In order to recover more heat, it is first necessary to identify the deficiencies of the device and alter its design to eliminate them. Many improvements are already at hand, though none are in commercial production. However, exchangers that incorporate efficiency features can be made by local metal-working shops at a cost of under $500, and the savings in fuel will almost certainly return that outlay within a few years.

The 1½-inch diameter tubes now used were selected because the material is a relatively cheap standard steel item and easily bent into a C shape. Steel tubing also has a long history of use in boilers and other heat-transfer equipment, so it's understandable that the first response of mass production was to use the familiar tube. But unlike most heat-transfer operations, in which heat is produced within an enclosed area and is thereby easily controlled, the tubular grate performs in an open fire that is difficult to control. And since air is a poor medium of exchange and picks up heat relatively slowly as it is blown through the tubes, most of the heat produced is being lost.

The round shape of the tubes is part of the problem, because at any given moment almost the entire volume of air lies in central areas where no heat transfer from the fire is taking place. The exchange of heat from fire to air occurs only at the tube walls and involves only *a film of air* making contact with them. In all other areas there is only an exchange of heat between cooler and hotter air molecules and *no increase in the total amount absorbed from the fire.* Of course turbulence is continually directing relatively cooler air to the tube walls, where it is heated, but this action merely displaces an equivalent amount of air, and the net effect remains the same. It is still only the film of air in contact with the metal walls of the tubes that is being heated by the fire.

Another facet of the same problem is that much of the heat absorbed by the tubes at any given moment is not conducted to the interior where it can be used but is lost by reradiation and convection to the firebox. The air inside the tubes (except for the film in contact with the walls) acts as an insulator and helps sustain the loss.

One of the problems in using air as a medium of exchange stems from the fact that oxygen and nitrogen molecules are "transparent" to radiant heat. They cannot absorb such energy, so the heat radiated inside the tubes by its walls does not cause the air to rise in temperature. In order to increase the efficiency of the fireplace, the air itself will have to be made a better medium of exchange.

If we redesign the grate heat exchanger so that a larger volume of air is brought into direct contact with heated metal surfaces, and if we also provide a means of conducting heat to the interior of the tubes from the walls, and convert the air into a better heat-transfer medium, the efficiency of the fireplace will be vastly increased.

Figure 26 shows an early attempt to increase heat transfer by inserting an eight-sided *metal vane* into each tube. (The grate is made of straight pipes joined by elbows to form a U rather than C shape.) Heat absorbed by the pipe walls is con-

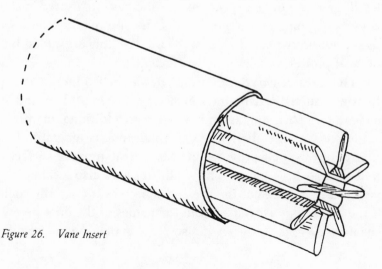

Figure 26. Vane Insert

tinuously conducted to vanes, and since steel is a far better heat-transfer material than air, the vanes attain fairly high temperatures. The heated metal provides additional surfaces from which air moving through the tubes absorbs heat and thereby increases efficiency 2 to 4 percent.

In a later refinement, the vane was enclosed in a thin steel sleeve slightly smaller in diameter than the tube. A metallic paste with good heat-transfer characteristics was spread over the sleeve before insertion to ensure continuous contact with inner pipe walls and better heat conduction to vanes.

Improved heat recovery due to vanes underlines a basic deficiency of the grate design: there is simply too little steel presented to the fire. The heat that each individual volume of air fails to amass in the second or two of its passage through the tubes is lost.

In an effort to increase the area of steel surface presented to the fire, an innovator reduced the gaps between tubes to $\frac{1}{2}$ inch and used the resulting space for more tubes. In another assembly, the upper tubes were installed without gaps between them, with the end ones separated from the firebox walls by a $\frac{3}{4}$-inch gap to allow evacuation of the exhaust. (A $\frac{3}{4}$-inch gap between lower tubes was left so that ash could fall.) The increase in efficiency was 3 to 6 percent—proportional, as one might expect, to the increase in heated metal surfaces. (In one design, steel deflectors slip over the ends of tubes and direct heated air downward. No btus are added but the heat produced is used more efficiently.)

The *3-shaped grate* in Figure 27 has a shelf 5 inches above the lower tubes that permits burning more logs. The unit is made in two parts and assembled with an additional manifold at the ends of the shelf to form a continuous air path. The tubing for the shelf provides increased heat-transfer surfaces, in itself a positive factor. The shelf tubes are also located in a particularly hot area of the fire and obtain heat from the coals of the upper logs as well as from the flames of the ones below. The burning logs on the shelf also provide the uppermost tub-

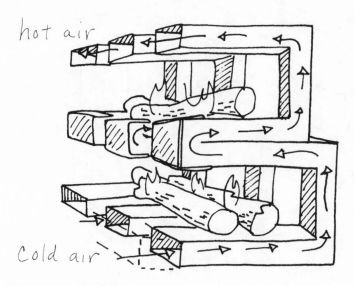

hot air

Cold air

Figure 27.   3-Shaped Grate

ing with a better source of heat than before, since there is now
direct contact with flames. The combined factors improve the
rate of heat recovery 2 to 4 percent.

Although well conceived, the shelf grate retains deficiencies caused by the round shape of the tubes. Curved surfaces
are poorer for heat transfer than flat ones (as we have all ex-
perienced when using a pot with a rounded bottom). Maximum heat transfer occurs only at the point of direct contact
with the heat source and becomes increasingly less efficient
the further the distance between the curved surface and heat
source. In addition, heat moves in a straight line, so only the
small portion that strikes the tubes at a right angle undergoes
a maximum exchange; most of the rest strikes the tubes at a
tangent, and heat transfer suffers accordingly. The top half of
the upper tubing receives much less heat than do areas in contact with coals or flames, and the same is true of the undersides of the lower tubes. Although the curved rear wall of the
firebox radiates some heat to the upper sections, as do small

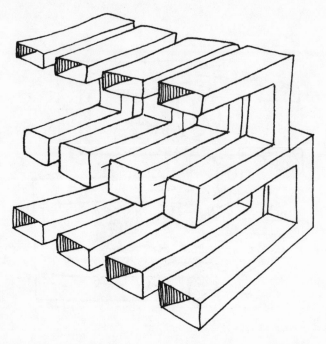

Figure 28. 3-Shaped Flat-Tube Grate

coals fallen through the grate to the lower, large areas of the grate are only partly involved in transferring heat. The condition is more pronounced at the side of the grate, where heat production and radiation from the firebox are least. At all points where heat exchange is taking place, the curved surfaces reduce heat recovery.

To eliminate the drawbacks of round tubes, another innovator flattened them as shown in Figure 28. A 1-inch hollow interior is left for air passage and, as in the round-tube version, halves are assembled with an additional manifold over the ends of the shelf.

In this design, a good part of the heat-absorbing surfaces lie at a right angle to the heat rising from the fire, the best position for heat transfer. Inside the flattened tubes, the distance air must travel before making contact with a heated surface is much shorter. In addition, heat radiated internally from the tube bottoms is transferred to upper and cooler sections more quickly, and a higher overall temperature results. The greater volume of air making contact with hotter surfaces more frequently produces a markedly better heat exchange.

Although the amount of air involved in heat transfer is increased in flat tubes, a significant part still lies outside the process at any given moment, wherever it fails to make contact with metal surfaces.

Wire lath, designed as shown in Figure 29, is a cheap material commonly used as a base for plaster. It is a continuous mesh of steel strands with $\frac{1}{4}$-inch diamond-shaped holes throughout. When used in conjunction with flat tubing, a greater amount of air in heat transfer at any given moment results. The tubes are filled with 1-inch-wide strips of lath stacked against one another so that heat is distributed by conduction while still permitting the passage of air. Heat absorbed by the outer surfaces of the flat tubes is conducted by the interlocking strands of metal throughout the interior. As air is blown through the holes in the mesh it encounters numerous heated surfaces, and while each strand is relatively small, their aggregate greatly enlarges the area of heat exchange. Heat radiated internally is also absorbed by the strands to improve efficiency further.

While lath causes more air to be involved in heat transfer, it does not change air from a poor medium to a better one. To solve this fundamental problem, we can look at the way solar radiation behaves in the atmosphere. As the sun's radiant heat enters the atmosphere, most of it passes undisturbed through

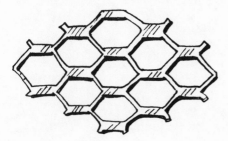

Figure 29.   Wire Lath

oxygen and nitrogen molecules, but a significant portion is absorbed by carbon dioxide, carbon monoxide, sulfur dioxide, dust, and, most importantly, water vapor. We can transform air into a much better heat-transfer medium by incorporating into it any of these molecules—most feasibly, of course, those of water vapor.

Figure 30 shows a *humidifier*, a simple arrangement for introducing water vapor into the air. A covered metal tray, partly filled with hot water, is installed next to the fan in the air path. Air blown over the hot water picks up vapor before it is delivered to the grate. Water can be heated before filling the humidifier at the beginning of each fire, or it can be heated by the fire itself in a variety of arrangements. One can also simply run the air through a humidifier before routing it to the fan.

Humid air is a much better heat-exchange medium than dry air, and when it is used in the grate a good deal more heat is transferred. The amount of heat is proportional to the humidity; the more water vapor in the air, the greater the amount of heat absorbed. Ideally, the air should be saturated with moisture, but of course this would be uncomfortable.

Figure 30. Humidifier

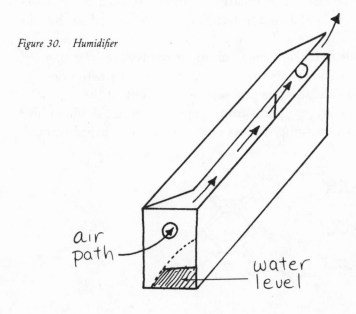

air path

water level

Comfort levels vary with individuals, but 35 percent is a common standard. The amount of humidity can easily be controlled by using hotter or cooler water or by increasing or decreasing the amount in the pan. One should simply experiment with the arrangement to see what is most comfortable.

The introduction of water vapor has a beneficial effect beyond improved heat transfer. Musical instruments and furniture, for example, are dried out and damaged due to "dry" heat each winter, and water vapor prevents this from happening.

The 3-shaped flat-tube grate filled internally with lath and supplied with humid air increases efficiency to between 20 and 25 percent. Although this represents a marked improvement, 75 to 80 percent of the heat is still being lost up the flue. The fundamental principle—that more heat is recovered as more heat-transfer surface is presented to the fire—points to further solutions.

As we look at the firebox, it is obvious that the back and side walls have the largest surfaces in contact with the fire and therefore offer promising locations for more heat-exchange surfaces. The sides are tapered and the back is sloped to radiate more heat into the room, but the effect is minimal; the contribution of these surfaces toward heating the room is about 1 or 2 percent.

Figure 31 shows a double-walled *lath-filled hollow steel jacket* designed for placement against the firebox sides and back. The jacket is filled with 1-inch strips of wire lath to conduct heat to the interior. It is used together with the grate and its fan; the air stream is split by a tee, half the air going to the grate and half going to the jacket via a manifold attached to the jacket inlet. The fire heats the steel sheets and the lath inside them, and the heat is transferred to moving air and then blown out the opposite side into the room. The large area of steel surface and the better heat-transfer characteristics of steel compared to firebrick give an increased efficiency of 11 to 15 percent.

The firebox, now clad with heat-exchange surfaces filled

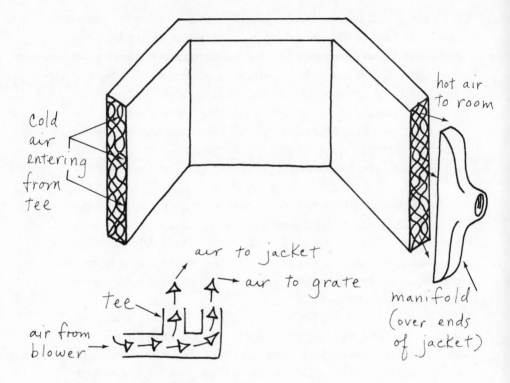

Figure 31. Lath-Filled Hollow Steel Jacket

with wire lath, and circulating humid air, returns 31 to 40 percent of the heat produced to the room. One additional gain can be made in the firebox by better control of the fire itself.

The greater a fire's supply of air, the more quickly the wood burns. Unlimited air not only causes the inconvenience of refueling the fire more often, but it also wastes heat. A much larger volume of air than is needed is drawn into the firebox from the room, heated by the fire, and escapes through the flue. The continuous presence of excess air also reduces the overall temperature in the firebox and leads to incomplete combustion of gases released by the wood. The smoke that is characteristic of open fires is visual evidence of incomplete combustion, but unignited gases that are not seen represent a far greater loss of heat.

In standard fireplaces, an adjustable metal plate called the *damper* is used to restrict the supply of air, and thereby control the rate of burning. However, most dampers, even when positioned as nearly shut as possible, leave a large opening that permits an excessive amount of air to be drawn into the fire. The damper is opened fully to start the fire, but when the fire is burning briskly and air should be restricted, it is hard to reach the damper to close it.

A better way to control the fire uses *glass doors* like those pictured in Figure 32. The glass is specially tempered for fireplace use. A small adjustable damper in the bottom rail provides an excellent means of regulating the amount of air admitted to the firebox. The fireplace damper is not needed now and is left fully open.

Glass doors have been on the market for many years.

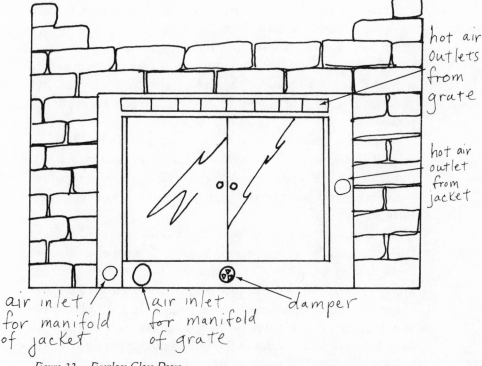

hot air outlets from grate

hot air outlet from jacket

air inlet for manifold of jacket

air inlet for manifold of grate

damper

Figure 32. Fireplace Glass Doors

They make for higher temperatures in the firebox, more complete burning of released gases, and better heat production. They eliminate the presence of excess air and attendant heat losses. The doors also provide a much more uniform rate of burning and, therefore, a more even performance by the heat-exchange devices in the firebox. Moreover, the problem of smoke being drawn into the room from the fire is eliminated.

The use of glass doors alone increases efficiency 3 to 5 percent. They are best employed as part of the heat-exchange system previously described. Many of the glass-door assemblies on the market have stationary rails at the top and holes to match the exhaust holes of round tubes used for the grate. If the more desirable flattened tubes are used, a custom rail has to be provided, but it will be well worth the additional expense.

When using glass doors, one opens the damper fully at the start of the fire and, when the fire is burning briskly, closes it to the smallest opening that permits the fire to continue burning briskly without producing smoke.

The glass loses its temper gradually when exposed to temperatures of 600 degrees or higher. The doors may last five years or twenty, but sooner or later they will have to be replaced.

The remaining factor to be considered in firebox heat recovery is the flow of air through the heat exchangers. Although the *blower* provides no heat, it is a fundamental part of the process and must be the right size if the system is to perform efficiently.

Let us assume the firebox contains steel-jacketed walls, a flat-tube grate heat exchanger, and glass doors. Let us also assume that we use a blower that delivers 10 cubic feet of air per minute (cfm), and that the air has a temperature of 70 degrees at the inlet and 130 degrees at the outlet. By multiplying the volume of air per minute, 10, by the temperature rise, 60, we have amassed in one minute the btu equivalent of 600 degrees to heat the room. If we now use a fan that delivers 100 cfm with an inlet temperature of 70 degrees and an outlet temperature of 100 degrees, although the outlet air temperature is lower than before, we get a total btu equivalent of 4,000 degrees, much higher than previously. Obviously, unless an optimum amount of air is delivered, all the projected increases in fireplace efficiency will suffer.

This optimum amount of air depends on the size and efficiency of the heat-exchange devices, the amount of heat produced by the fire at any given moment, and a number of other variables. A projection could be calculated beforehand for any one fireplace, but the result would still be a guess, and in these situations engineering practice usually comes down to pragmatic testing. The only instrument needed is a thermometer that will register temperatures between 60 and 170 degrees.

Before beginning the test, one should understand that the amount of heat produced by any fire varies continuously, particularly at the start and end. Ideally, the blower should respond by delivering the least air at the beginning and end, the most when the fire is at its peak, and in varying amounts at other times that correspond to actual heat production. One would need sensors and controls linked to a computer and a variable-speed blower to obtain accurate responses continuously, and this obviously is impractical.

While the blower to be used will deliver a constant amount of air, maximum heat transfer can be obtained by a unit of the correct capacity *when the fire is at its height*. Although this means that more air than is needed will be delivered at other times, heat recovery is not adversely affected. One is simply trading off the minor factor of fan efficiency in favor of the major consideration of maximum heat recovery.

To arrive at the optimum capacity of the fan for any situation, begin by installing one rated at 300 cfm. (Suppliers will usually lend you a succession of blowers if they believe you will end up buying one of the correct size.) A 300 cfm blower will deliver more air than you need unless your firebox is a good deal larger than the average dimensions of 36 by 26 by 22 inches. Measure the blower outlet hole and draw a circle the same size on a piece of cardboard. Divide the circle into eight equal parts and cut out one of the pie-shaped segments. Tape the cardboard over the outlet hole and, when the fire has attained full strength, measure the temperature of the air at the blower inlet and at a centrally located outlet of the grate. Let us assume that the cold air was 70 degrees and the hot air 180, a difference of 110. Multiply 110 by 37.5 (the restricted cfm of the blower) for a value of 4,125. Now cut out an additional segment and take the temperature again. Let us say the cold air is now 72 degrees and the hot air 174, a difference of 102. Multiply 102 by 75 (the new cfm) and 7,650 is obtained, a value higher than the first. Continue to remove segments and repeat taking temperatures until the value obtained is approximately the same as the previous one, neither significantly larger or smaller. That previous cfm is optimum. Multiply the number of segments removed by 37.5, the cfm value of the first segment, and the result is the cfm capacity that is most desirable.

Without altering the structure of the fireplace itself, by using a lath-filled flat-tube grate, lath-filled hollow steel jackets, and glass doors and by circulating moist air, we have elevated the efficiency of the fireplace to between 35 and 44 percent. Burning wood in this fireplace is a far cheaper source of heat than electricity and slightly less expensive than oil.

Cost-effectiveness compared to oil can be increased if the previously described pairs of vent openings are installed. The gain is not 5 to 10 percent any longer, since the improved fireplace has already extracted a good deal of the heat produced. But while the improvement is small, only 2 to 3 percent, the expense of providing the openings is also small. Since the advantage is permanent, it's worthwhile.

New and Altered Fireplaces

Although efficiency has been improved to a range of 35 to 44 percent, the rest of the fire's heat is still being lost up the flue. The use of new alloys with better heat-transfer coefficients could improve performance, but such future innovations, welcome as they will be, are not likely to produce substantially better efficiencies. The problem of major heat loss will remain, and if still better performance is to be achieved, the heat that goes through the damper will have to be utilized. This means, of course, altering the fireplace to gain access to the area above the firebox.

Reaching this area is no problem during construction, but if you already have a fireplace, you will probably regard the prospect of taking down part of a massive brick structure with a good deal of apprehension. Fireplace alteration is rare, and there is no body of experience or set procedure that would prove your fears unjustified. However, similar work occurs regularly. For example, all the techniques for obtaining access to the interior of a fireplace are also used to install a window in an existing brick wall, an infrequent but by no means extraordinary occurrence. The chimney is more stable than a single wall due to its four-sided construction, and removing a portion is simpler. A competent mason can perform the needed work using age-old, time-tested techniques.*

* There is no predicting what will be found behind the wall above the firebox. My experience shows that if the fireplace was built before 1960 there will be a void, and beyond that a smoke chamber wall. Until 1960, a general building code requiring the filling of such cavities with concrete was ignored by both building inspectors and masons, who considered the code nonsense (even in earthquake-prone regions, where they felt that the filled chimney might topple on the house in a solid mass rather than break into pieces outside). After 1960, inspectors began to insist that cavities be filled with concrete. It then became a usual practice for masons at the end of the day to dump broken concrete blocks and other noncombustible materials into the cavities and cover the debris with a thin layer of mortar they would have discarded anyway. This practice is still current, so finding solid concrete behind a chimney wall remains unlikely.

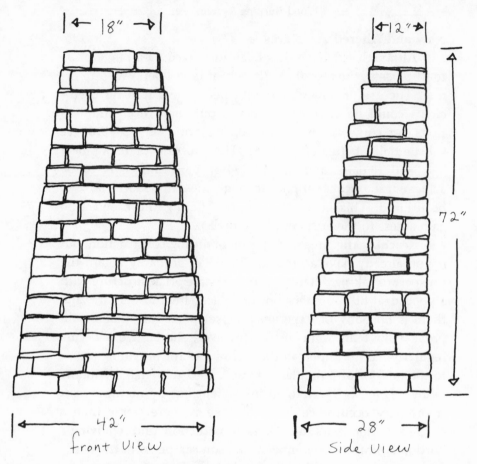

Figure 33. Smoke Chamber

Fireplace heat that has not been directed into the room passes through the damper and into the *smoke chamber* in the form of hot gases, steam, smoke, and thermal radiation.

The chamber is an enclosure of bricks within the chimney that keeps the fire's exhaust from the chimney walls. Its floor is comprised of the upper side of the damper, the upper side of the rear firebox wall, and a concrete-filled rectangular strip called the *smoke shelf*, which aligns with the *flue* and lies between the back of the firebox and the rear chimney wall. As can be seen in Figure 33, the smoke chamber is essentially an inverted funnel that collects the exhaust and focuses it toward the flue liner at its upper end. The chamber also provides structural support for the flue liner. Smoke-chamber dimensions vary with the likes of each mason (pictured is my own preference).

This smoke chamber is larger than what one might expect from the usual fireplace drawings. At its base, it is 40 inches wide and 28 inches deep, and it rises 72 inches from the top of the firebox. The rear wall is vertical and lies adjacent to the rear chimney wall; the remaining walls slant inward as they rise to a 12-by-18-inch rectangle, on top of which the flue liner is laid. The interior of the smoke chamber is four times larger than the firebox and offers an ideal location for heat recovery.

Figure 34 shows a standard *tube-jacket air-to-air heat exchanger*. Hot exhaust moves upward through tubes located within an enclosed cylindrical jacket, and heat from it is transferred to the outer walls of the tubes. Cold air enters the jacket through an inlet pipe at the bottom, circulates within the jacket and around the tubes, and leaves through an outlet pipe at the top. During this circulation, heat from the outer walls of the tubes is transferred to the cold air. The exhaust itself does not mix with the air to be heated; only its heat is transferred.

This standard heat exchanger is regularly used in boiler and furnace flues for capturing heat that would otherwise be lost, but its design and high cost make it unsuitable for capturing the heat that leaves through the fireplace damper. To be cost-effective for a fireplace, a heat exchanger should be located inside the smoke chamber where exhaust temperatures are highest, and it should present flat rather than rounded surfaces for heat transfer. The exchanger should also conform to the shape of the smoke chamber and occupy as much of its area as possible without reducing the draft excessively. Standard tube-jacket units do not fulfill these fundamental requirements (although they can work efficiently in a fireplace flue, as will be noted shortly).

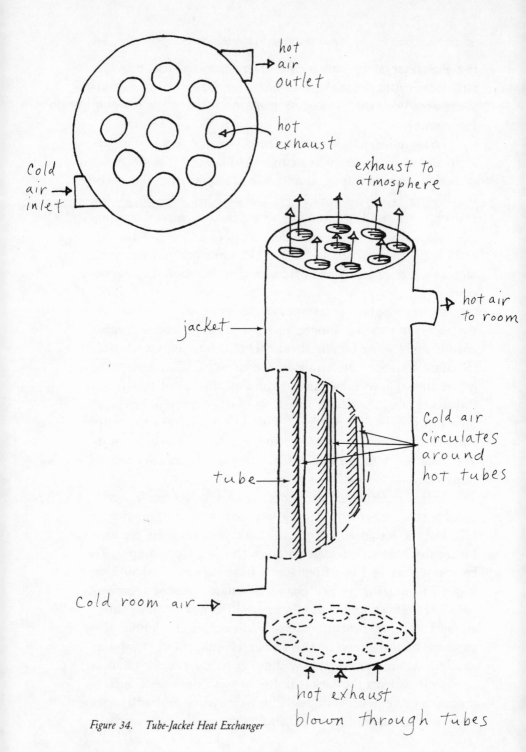

hot
air
outlet

hot
exhaust

Cold
air
inlet

exhaust to
atmosphere

hot air
to room

jacket

Cold air
Circulates
around
hot tubes

tube

Cold room air →

hot exhaust
blown through tubes

Figure 34.   Tube-Jacket Heat Exchanger

Figure 35 shows an experimental *seven-unit heat exchanger* positioned inside the smoke chamber. Each unit consists of a steel case filled with lath in which room air is circulated, plus an inlet and outlet pipe. (Construction is identical to the units used in the firebox.) The inlet and outlet pipes rest on the side walls of the smoke chamber; all units are linked together by common air-inlet and -outlet manifolds located in the space between the side chimney walls and side smoke chamber walls, as shown in Figure 36. They are slanted at various angles to best effect heat transfer from the exhaust without reducing the draft excessively. A blower at the chimney base, in the basement or crawl space, receives air through a duct and register in the floor at the end of the room opposite the fireplace. The room air is blown to the manifolds and through inlets, absorbs heat as it circulates through the interior of each unit, and is then blown back into the room through three outlet holes on each side of the fireplace.

The location and angle of the exchangers is a critical performance factor, for they must absorb as much heat as possible without blocking the draft too much. Their proper placement requires a closer look at the exhaust action of the fireplace.

There is a good deal of turbulence in the lower third of a burning firebox, as evidenced by the erratic upward movement of flames. On rising, the exhaust becomes increasingly influenced by suction created at the top of the flue, and turbulence diminishes markedly. The exhaust path straightens and increases its upward velocity, and the hot gases draw closer together as they approach the relatively small opening of the damper door. The damper, two or three bricks above the top of the firebox, isn't visible from the room, but on closer inspection one can usually see an accumulation of curling smoke waiting to pass through the constriction. Pressure for upward release is greatest at this point. The exhaust is also at its highest temperature here, and it follows that heat exchange should begin immediately beyond the damper door.

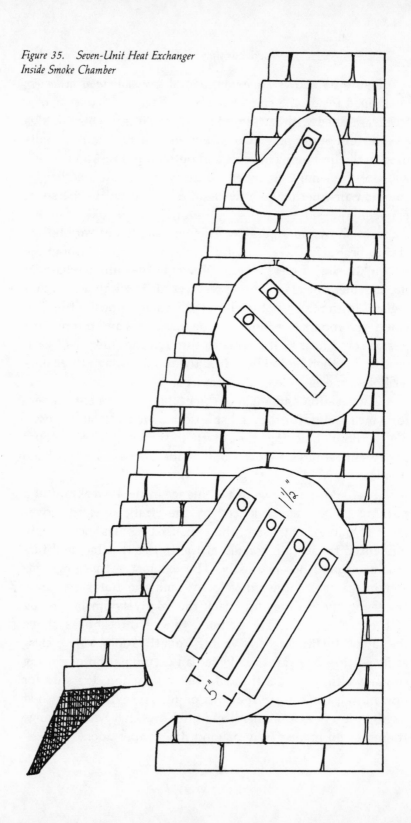

Figure 35. Seven-Unit Heat Exchanger
Inside Smoke Chamber

An obstruction in the path of the exhaust will diminish the force of the draft, and the placement of heat-exchange units must take this into account. Fortunately, modern fireplaces produce drafts a good deal stronger than needed to exhaust all the smoke, especially when the fire is intense. The draft can be reduced somewhat without reducing evacuation, but within limits. For example, if the first unit of exchangers is installed immediately behind the damper door and at a right angle to the exhaust path, heat transfer will be optimum, but too much turbulence will be created, lowering the draft to an ineffective level. Smoke will billow into the room or the fire will quickly be extinguished. The unit must therefore be placed in a *less than optimum* position (in terms of heat transfer) in order to maintain an adequate draft.

The velocity of the exhaust is another factor to consider in the placement of units. The velocity is not constant from the time the exhaust is produced until it is finally released into the atmosphere at the top of the flue. The movement is pulsating, and it slows at the constriction in the damper area. As the pressure from below and suction from above increase and overcome resistance, the exhaust shoots through the damper door forcefully.

When an old fireplace is demolished, there is always a pitted area in the rear wall of the smoke chamber 18 inches or so from the damper door that has been made by the hot exhaust. The area is higher than the damper, at an angle of about 15 degrees. This angle is the path the exhaust takes naturally as it passes through the damper door when the fire is hottest.

The four lowest exchangers are therefore tilted at a 15-degree angle for lowest velocity reduction. They are also positioned 1 inch apart to prevent excessive turbulence.

As the exhaust strikes the relatively thin leading edges of the units and splits into separate portions, its lateral force is diminished and its natural path altered. The continuous upward force exerted by the draft has moved the exhaust against six steel surfaces (the outer surfaces of the top and bottom units are outside the draft), and contact with them is maintained throughout their length.

The 1-inch space permits the installation of four units in the relatively small damper area. Such a space is sufficient for exhaust movement and also narrow enough for frequent contact of the exhaust with steel surfaces on either side.

The reduced velocity of the exhaust keeps it in the area of this first level of exchangers longer than would occur otherwise, and more heat is exchanged. The combination of six steel surfaces and hotter exhaust in longer contact results in good heat transfer.

Friction and turbulence slow the exhaust as it passes around the four units, but the initial momentum is still strong enough to drive it into a 12-by-36-inch rectangular space between them and the rear and side smoke-chamber walls. It is moving comparatively slowly at this point, but on entering the unobstructed space and a still-strong draft, it picks up speed again as it rises.

Heat extracted by the four lower units decreases the exhaust and irrevocably reduces its velocity, but the reduction is not critical because there is still a very large temperature difference between the exhaust and the atmosphere, as well as pressure from below. These factors cause the velocity of the exhaust to increase again (although it will never be as high as it was initially).

The lower temperature of the exhaust and the smaller available space higher in the smoke chamber create less favorable conditions for heat transfer. This negative aspect is partially offset by tilting the fifth heat exchanger at a 30-degree angle, more favorable in this position for heat transfer. This exchanger is a good deal smaller than the lower ones and won't weaken the draft excessively in spite of its more inclined angle. The sixth exchanger, smaller still and 1 inch above the fifth, is slanted at the same angle and continues the deflection of the exhaust toward the front chamber wall. Veto three remaining exchangers

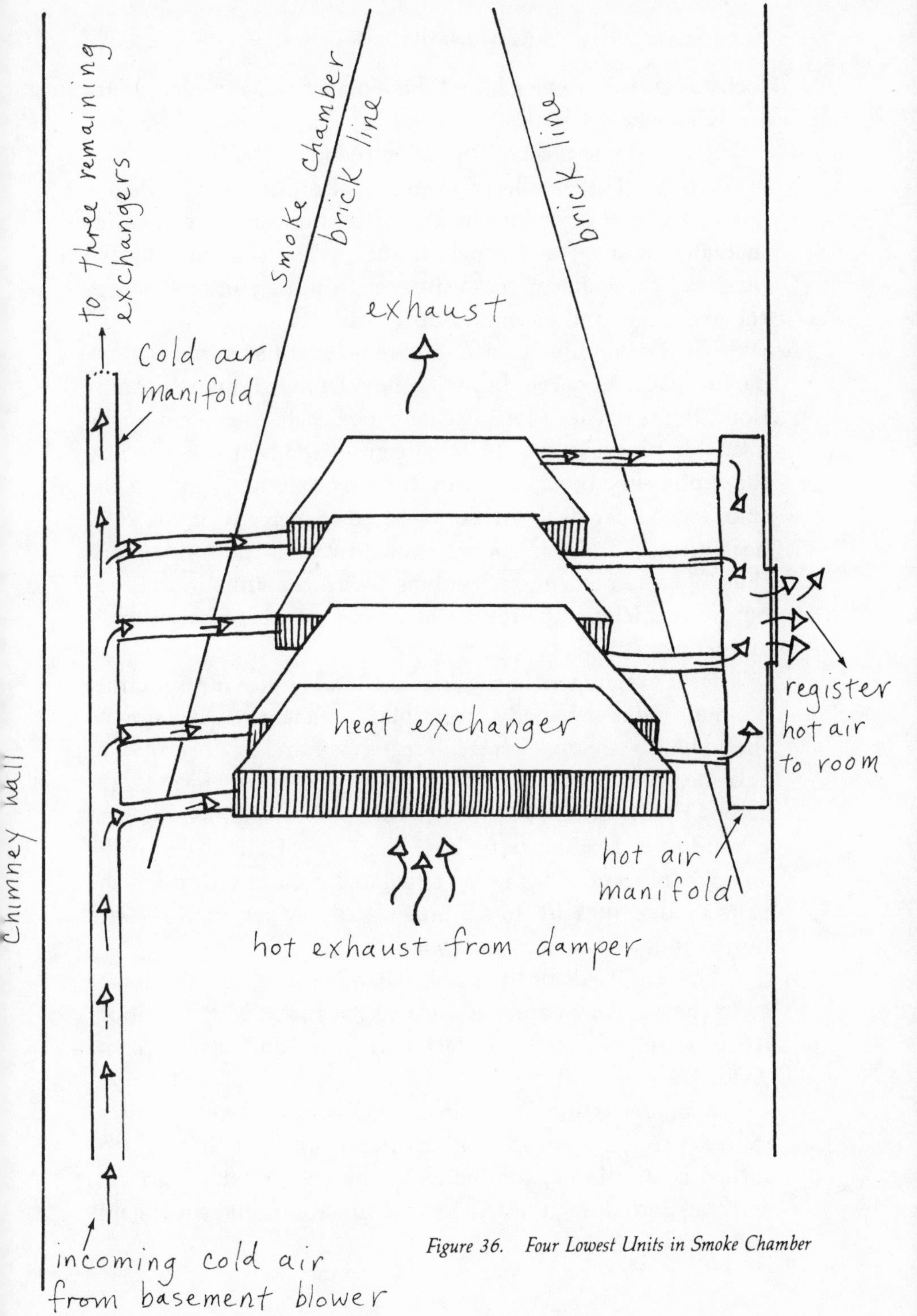

Cold air manifold

Smoke Chamber brick line

brick line

exhaust

Chimney wall

heat exchanger

register hot air to room

hot air manifold

hot exhaust from damper

incoming cold air from basement blower

Figure 36.   Four Lowest Units in Smoke Chamber

locity is further reduced, and the exhaust is again rising relatively slowly.

The exhaust reaches the seventh and final unit at its lowest velocity. This smallest exchanger, also tilted at a 30-degree
angle, has least effect on the draft. The unobstructed area immediately above it is strongly influenced by suction from the
flue, and the exhaust picks up speed once again as it leaves
the exchanger and continues upward.

The seven units have 90 square feet of steel surfaces compared to 30 square feet in the firebox. Under simulated conditions, the seven-unit heat exchanger extracts 35 percent of the
available heat. Since 35 to 44 percent of the heat produced by
the fire is already extracted by the firebox units, those in the
smoke chamber deal with only 56 to 65 percent of the total
heat produced. Their contribution to overall performance is
thus 19 to 23 percent, so combined with the units in the firebox, the efficiency of the fireplace is now up to between 54
and 67 percent.

Heat exchangers in the smoke chamber provide a number
of small indirect benefits in fireplace efficiency. Diversion of
the exhaust stream directs a greater volume against smoke-
chamber bricks than ordinarily occurs, so more heat is absorbed by the bricks. Since bricks are poor insulators, the heat
is conducted to their outer surfaces, part of which lie in the
area where room air is being circulated naturally through the
pairs of openings in the chimney walls. Much of this extra
heat is thus returned to the room.

The manifolds of the smoke chamber units are also located in the same area and add heat to the surfaces by radiation.
Room air makes direct contact with them and improves the
yield.

A similar action occurs inside the smoke chamber, where
the exchangers radiate heat continuously. The heat is absorbed by the brick, conducted to the brick's outer surfaces,
and transferred to circulating room air. Additional small gains

are provided due to the longer time the hot exhaust spends in the smoke chamber and the greater brick area to which it is exposed. The combination of these indirect benefits is approximately 2 percent.

The exhaust rising from the uppermost unit picks up a good deal of velocity as it enters the flue liner, still retaining a substantial amount of heat produced by the fire, and more heat can still be extracted. However, although the amount of heat left is large, the exhaust temperature has been reduced a great deal and the efficiency of an air-to-air exchanger will suffer. Since the flue liner measures 12 by 18 inches and extends only another 5 feet in a one-story house, there isn't much space left to make the transfer. In addition, the weakened draft in a one-story house will not permit further sizable reductions. However, if the chimney serves a two-story house and has another 14 feet of flue liner, the draft will be much stronger, and the process of heat extraction can be continued.

The tube-jacket heat exchanger, shown on page 96, is a standard air-to-air device that is widely available and suitable for use in the flue. Although it is not as efficient as other types, nor appropriate in the smoke chamber, it is manufactured in sizes that permit easy installation into the flue liner without impairing the draft too much. A 5-foot length, desirable in this situation, will recover an additional 3 percent of the overall heat produced and cost $320. The device is marginally cost-effective, and since the installation is permanent, will ultimately pay for itself.

The cost of a seven-unit heat exchanger installed in a new fireplace is $900. If the fireplace is already built, an additional $1,100 should be allowed for added labor costs. A completely modified existing fireplace incorporating all the improvements in the firebox and smoke chamber previously detailed will cost $2,450 (the optional tube-jacket heat exchanger in the flue is not included), and have an average efficiency of 65 percent. The cost for a new fireplace is $1,350.

Let's compare cost-effectiveness of the efficient fireplace and an oil furnace. The efficiency of furnaces is stated by manufacturers to be 65 to 70 percent, but in ordinary home use, due to frequent starts and lowered efficiencies at these times, overall efficiency is 55 percent. There is also an average yearly maintenance cost of $100 and an equal amount for depreciation. It costs the homeowner an average of 1¢ to obtain 1,000 btus.

If we burn dry hardwood in the fireplace and a cord costs $60, we'll obtain 3,033 btus per penny. There are no maintenance or operating costs (except for replacing the glass doors after many years).

Let us assume that your bill for oil in the past year was $1,000. Had you used wood exclusively, the bill would have been about $333, leaving $667 saved to amortize the initial investment in improving the fireplace. In the coming year the savings would be even more, assuming the price of oil continues to rise as expected. A new improved fireplace would be amortized in three years, at most. Amortization of an improved existing fireplace would occur in six years or less. Continuing to use the furnace for convenience would lengthen the time of amortization a bit, but the comfort of using it when most needed and eliminating all the hassles connected with the exclusive use of wood would certainly be worth it.

Building a new efficient fireplace or converting an existing one proves to be a highly cost-effective and practical alternative to oil-fired furnaces.

Stoves

After serving for more than a century as the main means of heating the home, and of cooking and obtaining hot water, the wood-burning stove entered the twentieth century as an endangered species. It avoided extinction in regions such as the Northeast where abundant firewood provided cheap, practical fuel. In the Catskills, where I live, for example, all the firewood needed for personal use can be cut without charge in a nearby state forest, and several homes in the area have been using wood continuously since colonial days. Some stoves have served many generations and continue in use. Of course, virtually all homeowners in temperate areas of the country switched to oil long ago, a situation that remained unchanged until the energy crisis in the early 1970s. Then came a dramatic resurgence of interest in wood heating; in the past year alone more than 300,000 stoves have been sold. A million homes today are heated exclusively by wood, another five million partially, and the projections for the coming years are even higher.

Modern stoves are essentially the same as those sold a hundred years ago and indeed are often manufactured with old equipment. They are made of sheet steel or cast iron, or a combination of the two. The wood is burned in an enclosed

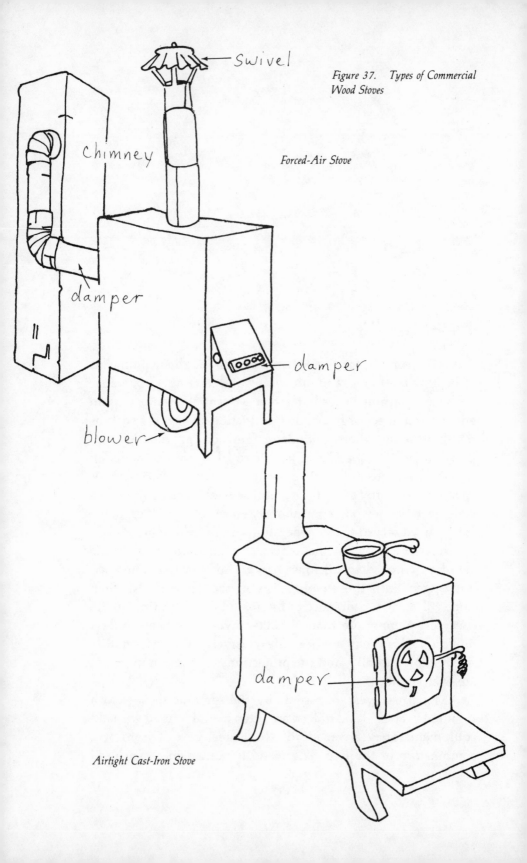

Swivel

Figure 37.   Types of Commercial
Wood Stoves

chimney

Forced-Air Stove

damper

damper

blower

damper

Airtight Cast-Iron Stove

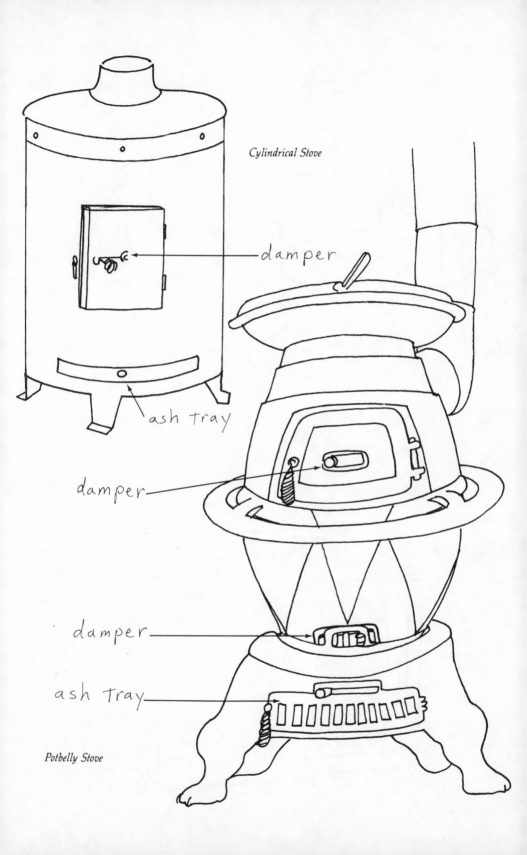

Cylindrical Stove

damper

ash tray

damper

damper

ash tray

Potbelly Stove

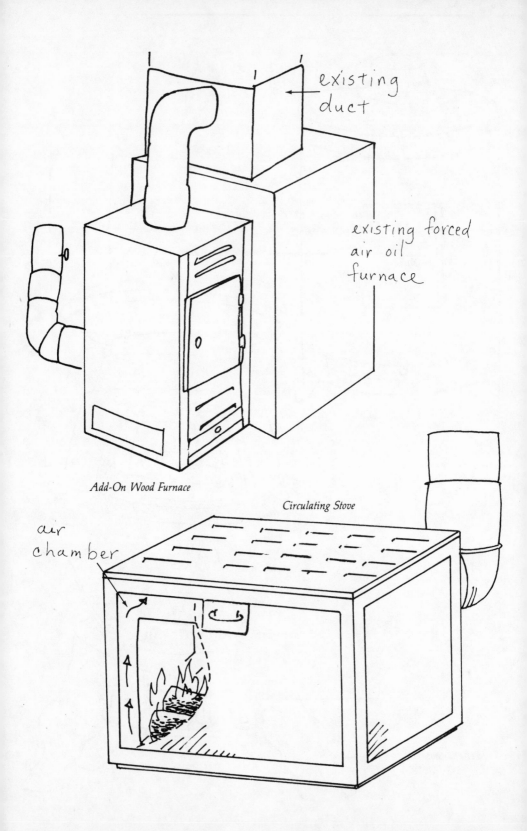

existing duct

existing forced air oil furnace

Add-On Wood Furnace

Circulating Stove

air chamber

firebox and heat is absorbed by the metal and radiated over the room. An adjustable *damper* (or dampers), either sliding or rotary, admits varying amounts of air and regulates the rate of burning. An elevated *grate* permits "primary" air to circulate below the wood and supply oxygen to the coals; "secondary" air above the wood supplies oxygen to burn released gases. The exhaust is vented to the outside through a *stovepipe*. This arrangement, developed largely through the last century, is extremely simple and continues to be the most widely used today.

Stove advertisements of the last century asserted efficiencies up to 100 percent, but having lived with a potbelly of this era and monitored its performance, I have found an efficiency of 30 percent closer to the truth. The few innovations established during the last century make the stove easier to deal with but do not deliver more heat to the room. So it's reasonable to assume that most any stove one buys today will also return around 30 percent. Even the scores of "new" stoves that have appeared in recent years with the boom in wood-burning—such as the airtight Scandinavian stoves—do not embody significant heat-transfer advances, and offer little increased efficiency.

A major cause of heat loss in a stove is the excessive admission of air into the firebox. Sheet-metal stoves are most prone to this deficiency. As the fire intensifies, the metal deforms and gaps are created around the firebox door, other hinged sections, and along poorly joined seams. A large volume of air is sucked in through the gaps, far more than is needed for combustion. The excess air absorbs heat and itself becomes a vehicle of heat loss as it is expelled into the outside atmosphere as part of the stove's exhaust.

An excess of air also causes a higher rate of burning. Sheet-metal stoves generally consume a full load of wood in less than an hour, entailing the extra inconvenience and expense of frequent reloading.

Many manufacturers equip sheet-metal stoves with cast-iron door assemblies. Cast iron maintains its shape at high temperatures much better than does sheet steel, and although the problem of gaps is not completely resolved, it is held to a minimum. With the cast-iron assembly, the stove damper can control the admission of air and the inconvenience of frequent reloading is eliminated. Sheet-metal stoves of this type are the cheapest available and perform almost as well as much more expensive all-cast-iron stoves.

If a sheet-metal stove is not equipped with a cast-iron door assembly, or if for any other reason an excessive amount of air enters the firebox with the damper fully closed, the condition is easily remedied by installing an additional damper in the stovepipe as shown in Figure 38. The section of stovepipe closest to the stove is removed simply by sliding it out and substituting a new section with a factory-installed damper. Constricting the stovepipe with the new damper reduces the draft, and less air is pulled through the gaps. Some unnecessary air will still enter the stove, but the amount is greatly reduced.

Ideally, the stove damper should be positioned at the

Figure 38. Stovepipe Damper

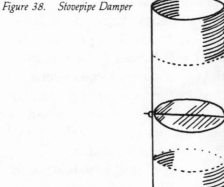

same level as the lowest logs and provide the sole source of air for the stove. It is only in this location that oxygen is needed. Dividing air into "primary" and "secondary" is a bit of stove-industry nonsense, for a single damper in the right location is perfectly capable of admitting the air needed to burn the gases of the wood as well as its coals. Not only do dampers for secondary air accomplish nothing, they are also often responsible for unwanted air admission. A single damper provides easier and simpler control of the fire and should be open only far enough to admit sufficient oxygen for combustion. Maintaining such control is not easy, however, for the intensity of the fire varies, requiring different amounts of air at different times, and unfortunately there is no sure mechanism to make such adjustments automatically. The best alternative is a *thermostatic damper* in the stovepipe, a poor substitute for automatic control but far better than continual tinkering with the damper. The device is activated by a bimetal helix that adjusts automatically to the temperature of the exhaust and thereby helps extend the life of the fire.

Incomplete combustion, another major source of heat loss, is difficult to control. About 40 percent of the potential heat in wood lies in the carbon monoxide, methane, and other combustible gases released by the wood and visible as flames. In sheet-metal stoves, a significant percentage of these gases is never ignited. Part of the problem is due to the flames being extinguished when they strike relatively cooler metal surfaces. The resulting smoke is evidence of incomplete combustion, a failure to convert all the gases to heat. The condition is aggravated by "cold spots" in the metal caused by erratic room-air contact with outer surfaces. Throughout the life of a fire in a sheet-metal stove, a significant amount of gases leave as smoke and invisible unignited exhaust that contribute nothing toward heating the room.

A major part of the problem is the relatively low operating temperature of the sheet-metal stove compared to cast iron. Sheet-metal stoves have a good deal smaller mass and are capable of containing a good deal less heat. Thus characterized by a relatively quick loss of heat, they deliver lower temperatures. A fire of 1,000 to 1,200 degrees is needed to ignite the combustible gases in wood, and in sheet-metal stoves these temperatures are reached only in the lower area of burning coal and flames. Once the gases escape unignited from this immediate area, there is nowhere else they can ignite, so combustion remains partial.

Cast-iron stoves also suffer from incomplete combustion, but to a lesser degree. Their greater mass needs a greater initial heat input before being brought up to operating temperature, but once that temperature has been attained, they function at several hundred degrees higher than their sheet-metal counterparts. This higher operating temperature not only creates a larger combustion area, it also enhances ignition, and a greater percentage of gases is burned than in the sheet-metal stove. Coals are burned more fully, as shown by a finer ash, and cold spots are eliminated. The evenness of heat flowing from the far larger mass of cast-iron stoves is more comfortable than the variable extremes of sheet-metal stoves.

Metal deformation in cast-iron stoves is minor and gaps are easily plugged. For example, the hollow V-shaped indentation often found around the firebox door can be partially filled with wet furnace cement. A small amount of oil is then rubbed over the corresponding V along the door perimeter. The door is shut and the cement allowed to dry. The result is a "custom" fit around the door that makes the assembly airtight.

Some stoves are lined with 1-inch-thick firebrick to increase operating temperatures and extend the life of the metal. Combustion is more complete, but less heat is transferred to the room due to the poorer conductivity of the brick. Thin firebrick is also fragile and can crack during ordinary usage. Stoves of this type are generally expensive and do not justify their additional cost.

Some stoves employ baffles (steel sheets) to route the exhaust through an S or similar shape in order to raise operating temperatures and reduce production of smoke. They are only marginally effective in this purpose and again do not justify the additional expense.

In the stove shown on page 107, baffles create a heated chamber between the firebox and outer walls. Room air is circulated in the cavities, but unless the circulation is forced, the increase in efficiency is too small to warrant the expense. In all of these more complicated and expensive stoves, one does end up burning the wood more completely, but only a very small amount of the heat gained is actually transferred to the room.

Some stoves of Scandinavian design employ a single baffle above the flames as pictured in Figure 39. Air is introduced at the front of the stove so that logs burn from front to back rather than all at once. A slower and more even rate of burning results and the interval between refueling is considerably extended. "Scandinavian" stoves are more convenient to use for these reasons, but they still do not increase the recovery of heat. They are also the most expensive type of stove you can

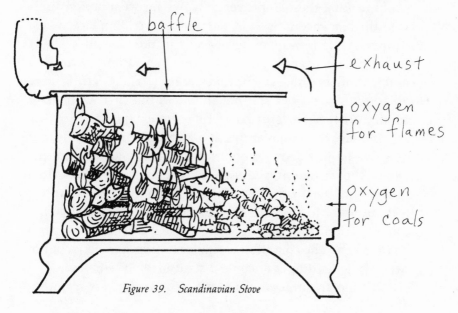

Figure 39. Scandinavian Stove

buy. (Brand names should be avoided. The stoves are actually manufactured in Ireland, and identical "nameless" versions are available at one-third the cost.)

Although there isn't a great deal of difference in performance between sheet-metal and cast-iron stoves, the latter is a better choice because of its more airtight construction. Still, the amount of heat lost in either case remains around 70 percent.

Among the hundreds of commercially manufactured stoves currently on the market, the most efficient is the forced air-circulator type shown on page 106. Steel baffles along the sides, back, and top isolate the firebox from the outer walls. Room air lying at floor level is circulated in the cavities between the baffles and outer walls, and the heated air is released to the room through a grille at the top. The convective gain compared to other stoves is about 5 percent, so 65 percent of the heat produced is still lost up the flue. Yet this is the best one can expect from a commercial stove. If further efficiencies are to be made, new and better means of heat transfer have to be introduced.

In making these improvements, we face a situation similar to the one encountered in the unimproved fireplace. There is no miraculous way of resolving the problem in a single stroke; greater heat recovery can be achieved only through a combination of modest improvements. (I am referring here just to the stove itself and not the flue.) But first we have to identify other heat losses not yet mentioned.

The highest temperatures and greatest amount of heat in the stove lie just above the burning wood. A glance into the firebox shows that this area of maximum heat makes little contact with steel surfaces, the sole medium of heat transfer to the room. Most of the heat absorbed by the metal comes from radiation from the fire, and not from direct contact with the fire itself. The condition is similar to trying to boil a pot of water while holding it at some distance from the flames.

The metal surfaces of a stove also obtain heat from hot

gases, but such exhaust temperatures are lower than those of flames. The hottest gases are concentrated around the central axis of the fire, in the center of the stove, and only relatively cooler gases make contact with the metal. If we could stop the process of burning momentarily, we could see that the hot gases have assumed the shape of the stove. *Only a film along the perimeter is actually involved in heat transfer.* The remainder of the gases, virtually the entire volume, does not contact the metal, and therefore remains uninvolved in heat transfer. (Turbulence in the firebox directs "new" gases to the metal, but this merely displaces one film with another and the net result is the same.)

Hot gases that are turbulent around the flames and immediately above them straighten quickly as they rise toward the flue. (We can see this by following the smoke.) From the time they are produced until they enter the flue, only a very small portion of these hot gases makes contact with metal. To increase the stove's recovery of heat, these hot gases must be made to contribute to heating the room.

The firebox of the stove is much smaller than that of the fireplace, so the space available for heat transfer is itself a good deal less. Such a relatively small area creates a delicate situation in respect to the fire. Any negative development in any portion of the firebox, such as excessive smoke production, influences the entire fire and can cause incomplete combustion. In redesigning the stove, this situation must be kept in mind.

Another liability of current stove design is the small space for wood in the firebox. A stove, unlike a fireplace, has no large chamber away from the fire where smoke can accumulate when the draft is weak. Smoke that cannot escape through the flue has a quenching effect on the fire, causing increasingly poorer combustion and even more smoke.

The draft of a stove is fundamentally important to heat transfer, and there are several factors involved. A stove flue is usually 5 to 7 inches in diameter, compared to the 12-by-18-

inch flue of a fireplace, and produces a comparatively weaker draft. Then too, stovepipe loses heat more quickly than the terra-cotta flue liner enclosed in a masonry chimney, particularly after it has left the interior of the house; and lower exhaust temperatures produce a weaker draft. Since the strength of a draft is related to the height of the flue, the shorter stovepipe, rarely extending above the height of a chimney, is still another negative factor.

If we are to improve the efficiency of the stove, more heat must be directed to the room—with, of course, a consequently lower temperature at the exhaust. It is therefore essential that a strong draft be present at the outset.

The popularity of wood stoves lies for the most part in their modest cost, durability, and simplicity of installation. Stoves offer an immediate alternative to the oil furnace, even though 30 to 35 percent heat recovery is hardly a model of efficiency, and improvements should not be made at the expense of the positive factors that made the stove attractive in the first place. Although none of the stove designs to be presented in the following section are yet commercially available, all take those factors into account. They have been fabricated in various localities at reasonable costs, and many have been successfully used for a number of years.

Improving the Efficiency of the Stove

Wood stoves are designed primarily for ease in manufacture, and their shapes have little value in heat transfer. If heat recovery were a major consideration, we would long ago have seen more suitable shapes, and most of them would have included finned surfaces. Finned surfaces similar to those on hot-water radiators have long been known to improve heat transfer, and their application to the surfaces of a conventional stove is easily accomplished, as indicated in Figure 40. Any sheet-metal shop can install fins by cutting slots in the walls of the stove and welding the fins with half their surfaces lying inside the stove and half outside.

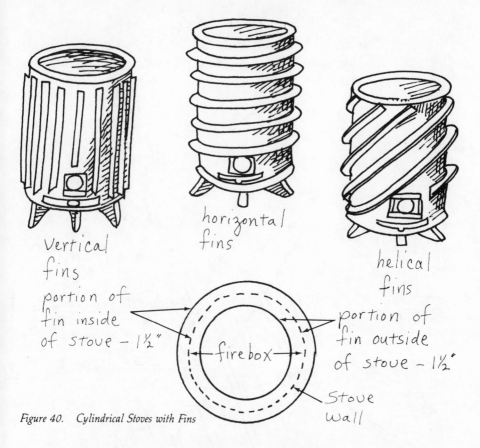

Vertical fins

horizontal fins

helical fins

portion of fin inside of stove – 1½"

portion of fin outside of stove – 1½"

firebox

Stove Wall

Figure 40.   Cylindrical Stoves with Fins

Heat produced by the fire is absorbed by the surfaces of the fins that lie within the stove, conducted to surfaces that lie outside the stove, and transferred to the room via radiation and conduction. The more closely spaced the fins, the better the heat recovery. Horizontal fins are least effective because heat accumulates along their undersides and is blocked from its natural upward movement away from the stove. Vertical fins, on the other hand, while they do not interfere with heat movement, provide little absorptive surface. Helical (spiral) fins offer the greatest surface without restricting the flow of heat, and are therefore most suitable. (The fins are usually steel, a moderately good heat-transfer material, but copper alloys are better conductors and improve performance.)

Fins improve a stove's efficiency in two ways: with addi-

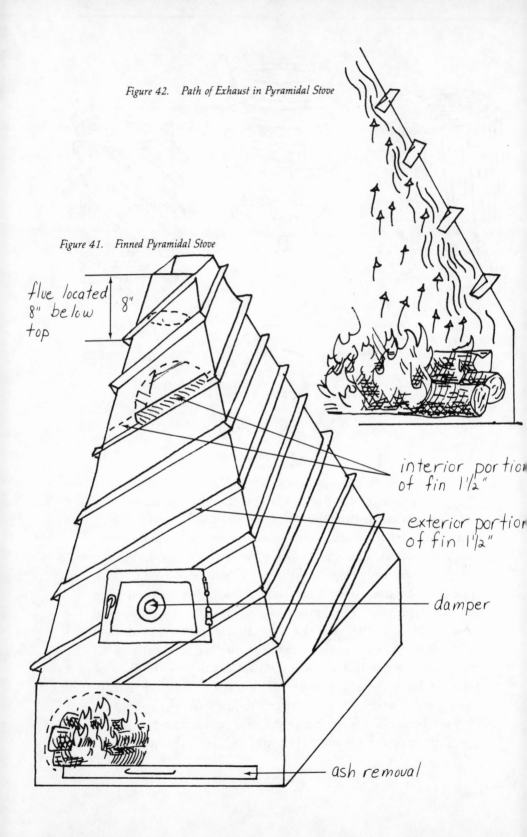

Figure 42. Path of Exhaust in Pyramidal Stove

Figure 41. Finned Pyramidal Stove

flue located 8" below top

8"

interior portion of fin 1½"

exterior portion of fin 1½"

damper

ash removal

tional heat absorptive surfaces inside the stove and additional surfaces on the outside of the stove for heat transfer to the room.

The *pyramidal stove* in Figure 41 ensures good contact of heat and metal. The shape of the stove forces the exhaust into a path along the metal sides, as shown in Figure 42. In the lower section, flames strike metal surfaces directly, and in the upper, hot gases are in continuous contact with metal walls all the way to the flue. Except for a small portion of gases in the central area, heat produced by the fire is totally involved in the transfer and produces a much greater efficiency.

Figure 43.   Smoke Chamber and Heat-Transfer Areas

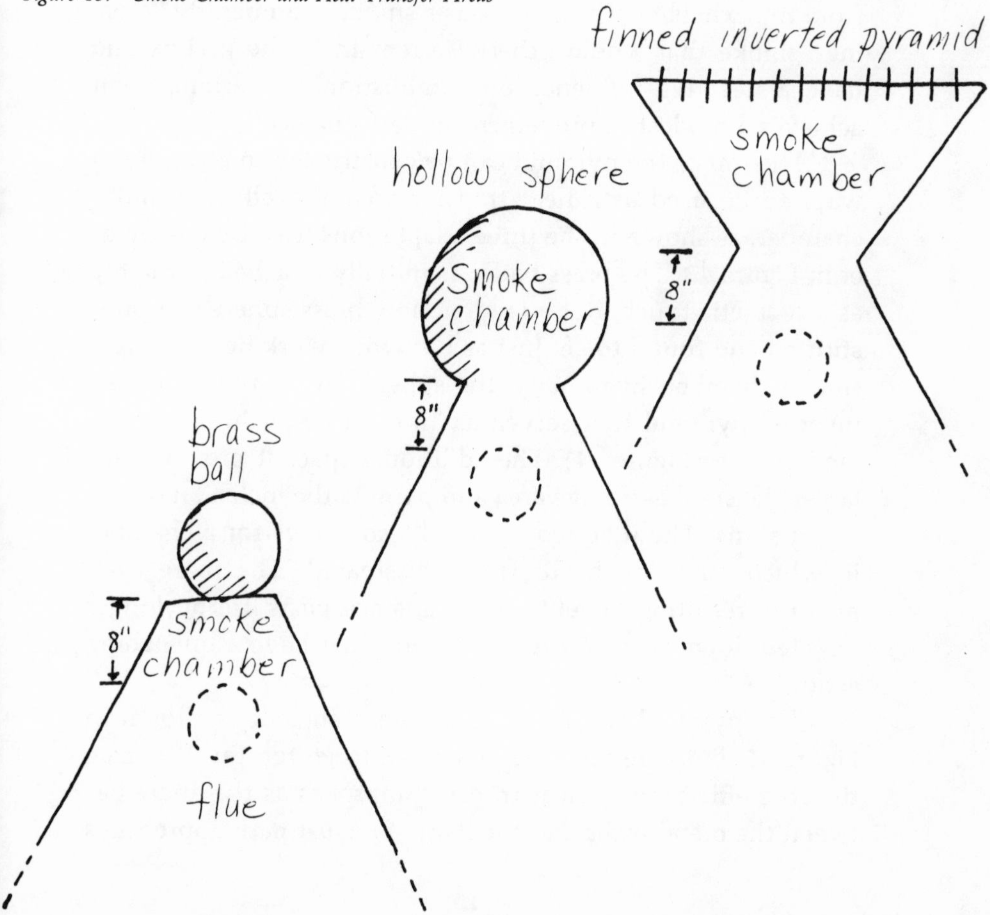

finned inverted pyramid

smoke chamber

8"

hollow sphere

Smoke chamber

8"

brass ball

Smoke chamber

8"

flue

The strongest draft is created when the flue is centered directly above the firebox. Medieval fireplaces were built this way, and if one looks up the throat of such a fireplace (there is no damper), the sky is visible at the top of the chimney (the chimney itself is the flue). Firebox and chimney are in the same vertical line. During colonial times, the throat of the fireplace was shifted toward the room, forcing the exhaust path into the shape of a V on its side. A smoke shelf was created, which largely eliminated downdrafts, but resulted in a weaker draft up from the firebox. A similar tradeoff has been made in the pyramidal stove. The flue opening lies on the rear wall 8 inches below the top of the pyramid. The placement results in a weakened draft but creates an area at the top of the pyramid, away from the exhaust path, to serve as a smoke chamber. By "storing" smoke that would otherwise remain in the firebox and have a negative influence on combustion, the arrangement achieves a modest improvement in performance.

The top of the pyramid can be constructed in a variety of ways to be used as a heat-transfer area as well as a smoke chamber, as shown in the three adaptations tried by one owner in Figure 43. The brass ball was initially attached to the top as a cosmetic touch, but a large hollow brass sphere was substituted and found to get just as hot while working better as a smoke chamber. Eventually, the sphere gave way to a finned inverted pyramid that served as an even better heat-transfer medium (see Figure 44). The additional space it provides enlarges the smoke-storage area and permits the inclusion of numerous fins. The inverted pyramid also allows for a chamber in which gusts in the draft are dissipated. The more even pressure resulting benefits the fire, since gusts are no longer directed downward to the fire where they have a quenching action.

The pyramid can be more severely angled, as shown in Figure 45, but care must be taken not to go too far. It is true that the efficiency of heat transfer improves as the angle between the metal walls and the flame-exhaust path approaches

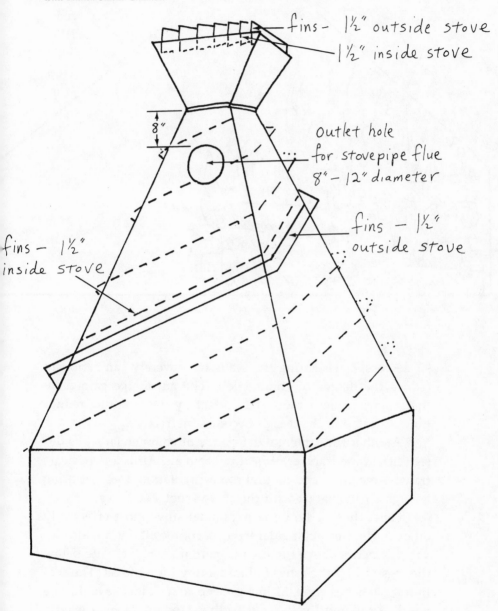

Figure 44. Rear View of Stove with Finned Smoke Chamber

fins – 1½" outside stove

1½" inside stove

outlet hole for stovepipe flue 8" – 12" diameter

fins – 1½" outside stove

fins – 1½" inside stove

8"

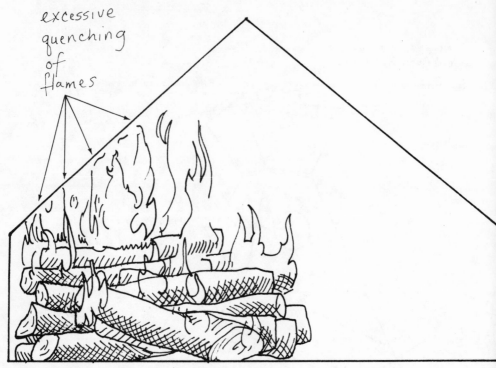

excessive quenching of flames

Figure 45.   Severely Angled Pyramidal Stove

90 degrees, but slanting the walls too severely can seriously reduce the degree of combustion. The advantage gained by the steeper angle is more than offset by the resulting reduction of heat, and the design becomes ineffective.

Another fundamental deficiency of commercial stove design lies in the transfer of heat to room air. Air is a poor heat-transfer medium, and natural room circulation does not bring enough air in contact with the stove's metal surfaces.

Here, then, is how the pyramidal stove most effectively takes shape: the rectangular lower section contains a grate elevated 3 inches above the floor to permit air to circulate below the logs and let the ashes fall into a tray for removal. The section is 30 inches wide, 22 inches deep, and 12 inches high. The width is tailored for logs cut to the standard 24-inch length, leaving enough space at each end for air circulation. The depth permits loading up to five logs on the grate and two or three additional courses above. The distance from the floor of the stove to the top of the pyramid is 72 inches, a height that provides both an optimum slope and a much better proportion of space to wood than is found in traditional stoves. The problem of air circulation can be resolved by a blower located at the back of the stove. Two hundred cubic feet of room air per minute is directed over the walls of the stove and back into the room. This large amount of cooler air reduces the operating temperatures in the firebox and decreases the combustion of gases, but the greater amount of heat delivered to the room offsets this loss. And the limitation of air as a heat-transfer medium can itself be improved by routing room air through a humidifier before it is drawn into the blower. This also makes a significant contribution to better heat recovery.

The combination of shape and other features described so far provides the pyramidal stove with an efficiency of 50 to 55 percent.

About 40 to 50 pounds of wood can be loaded into the stove to burn for six hours or longer. Since 45 pounds of wood produces 337,500 btus, we can estimate that at an efficiency of 52.5 percent, the pyramidal stove yields 177,000 or so btus to the room. As the stove is airtight, the rate of burning can be regulated simply by opening or closing the rotary damper. For example, when the temperature outdoors is zero degrees, a 1,500-square-foot house needs approximately 30,000 btus per hour to maintain an interior temperature of 70 degrees. We can hold that desired temperature by adjusting the damper so that the yield is spread over a bit less than six hours. On most winter days only about 10,000 btus an hour are required, and burning can be protracted over a longer period. When the life of a fire can be spread out to as much as an eight-hour period, the major inconvenience of starting a new fire in the morning is eliminated, and the practicality of using the stove improved. (I am assuming that the house doesn't have large expanses of thermally unprotected glass or other sources of major heat loss, and that the heat produced by the stove will be moved mechanically to all areas.)

By enlarging the stove to burn 75 to 100 pounds of wood, larger homes can be heated exclusively with wood and intervals between refueling extended to twelve hours and longer. However, stoves this large present the homeowner with the problem of having a heat source dominate a room used for living purposes. There are several options: we can retain the essentials of the pyramidal stove but alter its appearance so that its larger presence is more acceptable; we can increase the efficiency and retain the smaller size; or we can move the stove to the basement and call it a furnace.

Wood Furnaces

If we locate a stove that burns 75 to 100 pounds of wood in a basement, a duct-and-blower system becomes essential to deliver the heat to the house. For all practical purposes we now have a wood furnace. Traditional wood furnaces are still on the market and are usually connected to existing oil furnaces as shown on page 108. However, wood furnaces cost as much as oil furnaces and their efficiency is on the order of 40 percent (a bit higher than that of the best commercial wood stoves). This high initial cost and poor efficiency argue against their use. The larger pyramidal stove is a good deal less expensive, much more efficient, and easily adapted to furnace use.

Figure 46 shows the pyramidal stove converted to a furnace. Apart from its larger size, it remains essentially the same as the stove with the exception of a flat-tube, lath-filled heat-exchanger grate introduced into the firebox (see pages 84–86). The furnace blower supplies air to the inlet manifold. The air circulates through the lath-filled tubes and is then blown to the plenum of the existing oil furnace, from where it is ducted throughout the house in the usual manner. The exhaust is routed to a separate flue.

The use of a heat exchanger in the firebox is more effi-

Figure 46.   Finned Pyramidal
Furnace with Heat Exchanger

blower

damper→

air-inlet manifold

ashtray

cient than blowing air over the stove, and heat recovery is raised to between 55 and 65 percent.

Efficiency can be further improved to help heat the basement and the rooms above by capturing heat from an *extended flue* as shown in Figure 47. Instead of locating the pyramid directly in front of the chimney, as would be the usual practice in installing a stove, it is located at the opposite end of the basement and connected to the chimney flue by extended stovepipe. The stovepipe then becomes a heat exchanger by radiation and air striking its hot surfaces. The only limitation in extending the stovepipe is excessive reduction of the draft. But as long as the draft remains strong enough to evacuate the exhaust, *this is the simplest, cheapest, and most efficient method of heat recovery.* (We could, in fact, ignore efficiency of the furnace,

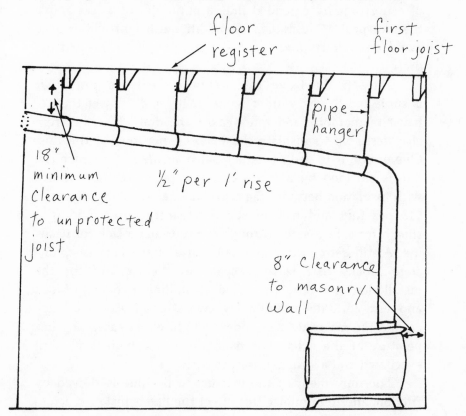

Figure 47. Flue Extended Across Basement

shape it to a simple steel box with a cast-iron door, provide a damper and outlet hole for the flue, and simply use the stovepipe to obtain maximum efficiency.)

To maintain the strong draft essential to recovering the maximum amount of heat produced by using the stovepipe as a heat exchanger, the exhaust outlet should have a diameter of 12 inches, instead of the 8-inch diameter ordinarily used for a wood furnace. The larger size creates a stronger draft by diminishing the effect of friction and providing a larger heat-exchange surface.

Ordinarily, the flue of a wood furnace is extended a foot or two above the roof line, and in better installations 2 feet above the peak, to avert downdrafts and back-puffing caused by wind swirling along the roof. Because of the weaker draft of a horizontally extended flue, a minimum of 4 feet above the roof peak is desirable. The additional height will at once avoid downdrafts and increase the strength of the exhaust draft.

Once the flue leaves the interior of the house structure it is subjected to cold winter winds that rapidly lower the exhaust temperature and can weaken the draft. It is therefore also desirable to insulate the exposed portions of the flue. One option is to use a product called *metalbestos*, in which the flue is shielded by a larger diameter pipe over the outside with insulation between the two. The material currently costs $13 per foot and has an added advantage of providing a sheath for safe passage through the roof and a jack for flashing. A still better alternative is the use of specially-cast concrete blocks that leave a space in their center for the installation of a terra-cotta flue. (Installing a concrete-block and terra-cotta flue is half a day's work for a mason and helper, and the materials cost under $50.) In either case, the temperature of the exhaust is maintained at a high level until evacuated and a stronger draft results.

The movement of the exhaust in the flue is slowed by friction, and to minimize this effect the flue is installed across the basement on a rise rather than horizontally. A rise of $\frac{1}{2}$ inch per foot is satisfactory. As a rule of thumb, the lateral distance should not exceed one-third the vertical distance (another reason why a higher flue is desirable).

One naturally wants to introduce the greatest amount of stovepipe in order to recover the greatest amount of heat. Theoretically all the heat produced should be available for recovery except the amount needed to drive the exhaust, but in practice friction and turbulence inside the flue cause a loss of draft before a loss of heat produces the same effect. (Swirling winds blowing about the flue outlet and down into it, even with the higher outlet, also contribute to the loss of draft.) Ideally, a strong steady wind should blow at a right angle to the top of the flue to create suction and assist the draft, but of course such action is a whim of the weather (unless a blower is positioned to do it).

Generally speaking, the exhaust temperature should not be lower than 275 degrees. But adequate drafts have been obtained at far lower temperatures, and with so many variables to account for the discrepancy, it's best to determine the optimum amount of stovepipe by pragmatic testing.

Figure 48. Pipe Section with Elbow and Tee

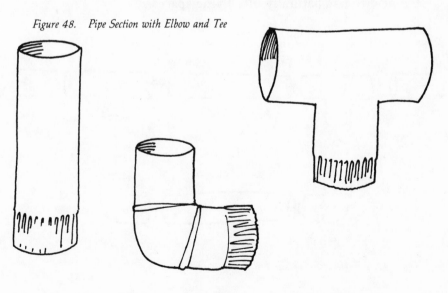

Figure 48 shows the stovepipe *elbow and tee* needed for determining optimum stovepipe length and making the actual installation. Stovepipe sections are slightly smaller in diameter at one end than the other; the smaller end is simply inserted into the larger end of another section or fitting. All items are standard and available at sheet-metal shops and lumberyards at a very modest cost.

The stovepipe is first installed with two tees 3 feet apart at the beginning of the run—no closer than 24 inches from a nonfireproofed ceiling and 18 inches below fireproofed ceilings—and is connected from the furnace or stove to the chimney flue. A parallel 3-foot section of stovepipe is then attached to the run with elbows as shown in Figure 49. If the draft permits, the second tee is repositioned 6 feet from the first and the parallel flue lengthened with another section of stovepipe. In the same fashion, the parallel flue is extended until there is a buildup of smoke in the firebox that indicates an overly reduced draft.

Too much heat may be radiated by the furnace and flue into the basement. Cutting holes in the basement ceiling and installing registers controlled from the floor above will enable the heat to rise naturally into living spaces.

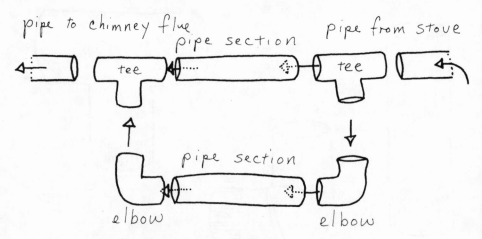

Figure 49.   Single-Loop Stovepipe Installation

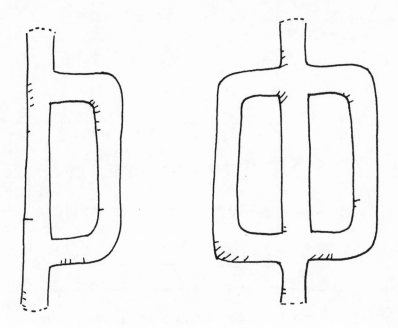

Figure 50. Single- and Double-Loop Flues

A long run of flue is desirable in the basement since the heat it yields will be spread over a maximum area. However, if the flue is in a living area, an aesthetic problem can arise. The obtrusiveness of the stovepipe can be reduced by using parallel vertical flues in a *single or double loop*, as shown in Figure 50. In this arrangement, the stovepipe can be concealed in the area behind the stove.

If we wish to further reduce the amount of flue for aesthetic reasons, but nonetheless increase heat recovery, a fan can be installed behind the stove to direct room air over the stovepipe and stove (see Figure 51). The same flue opening from the stove is used, the same opening to the chimney, and without altering the existing stove in any manner heat recovery is substantially increased.

We can remove the stovepipe completely from sight, if so desired, by placing a perforated heat shield in front of it and

Figure 51. Fan Behind Stove

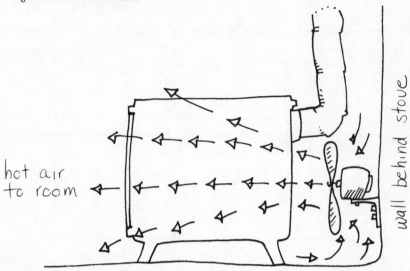

hot air
to room

wall behind stove

blowing air across the flue and through the perforations. By attaching helical sheet-metal fins to the stovepipe, heat transfer from the interior to exterior will be bettered and the amount of stovepipe further reduced. But no matter which arrangement is selected, extracting the maximum amount of heat produced is realizable in any stove through the use of an extended flue and forced air. The great advantage of using the flue as a heat exchanger is that it lends itself to *all stoves*, and this marked improvement in efficiency is obtained at a cost of less than $25. Perfect efficiency isn't possible, even with a very strong initial draft, but 70 percent or more can be realized in a traditional stove by converting the flue to a heat exchanger.

A stove with a 70 percent efficiency and a 50-pound load of wood yields 262,500 btus. Assuming an hourly requirement of 30,000 btus to maintain the indoor temperature of a 1,500-square-foot house at 70 degrees, a single load of wood burned at a relatively slow rate in a flue-lengthened stove can provide heat for nine hours or so. In the case of a traditional commercial stove with its restricted firebox area and its poor proportion of space above the fire, a load of wood will most likely be

132

consumed in six hours or less. However, most winter days will not require 30,000 btus, and if one has a stove-flue-blower arrangement, along with a 50-pound capacity, enough heat will be recovered to fulfill demand for eight hours or longer. Given the much higher efficiency of the pyramidal stove, the life of the fire can be extended to twelve hours or longer, with more than enough heat recovered each hour.

Summation

The optimum stove installation is a finned pyramidal furnace with a 75-to-100-pound capacity. The interior is equipped with a flat-tube lath-filled grate heat exchanger and air supplied by the existing oil furnace blower. Heated air is returned to the ductwork and distributed by the existing system. The flue is extended over the entire basement, and registers in the ceiling permit heat passage to higher areas. The oil furnace is used when automatic control is needed, such as when leaving the house on weekends, and the wood furnace at all other times. The two systems used together eliminate the drawbacks of exclusive wood use and provide heat at approximately one-third the cost of an oil furnace.

If the house has no basement, or if a wood furnace cannot be used for other reasons, the pyramidal stove with grate heat exchanger shown in Figure 46 is the next best choice. It will burn 50 pounds of wood, and the interval between refueling can be extended to eight hours or longer. It is only slightly less efficient than the furnace counterpart but is still highly cost-effective compared to an oil furnace.

If an existing stove is to be kept in use, there are several ways it can be made more efficient. A blower can be positioned behind it to provide forced air contact with metal surfaces and to blow the heated air into the room. As a second step, installing parallel flues will increase its heat recovery. And if the stove has a capacity of 50 pounds or more, a flat-tube heat exchanger should be introduced. If the capacity is less than 50 pounds, there is simply not enough space in the firebox for additional heat-transfer material and the stove should be regarded as only a supplemental source of heat for the room.

The fabrication of stoves, heat exchangers, and all other improvements in heat recovery detailed in this chapter should present no problems to any competent metalworker. Costs are modest; many "custom" stoves are regularly manufactured locally for $200 to $300. An improved stove or furnace is an immediate practical alternative or supplement to the oil furnace, and its greater use will not only help the environmental-home-energy picture, but also produce substantial savings for the homeowner.

⦚ Stacking and Utilizing Firewood

In a pyramidal stove or furnace, logs approximately 4 inches in diameter should be used, either whole or split (in smaller stoves, it's best to use split logs).

Logs should be stacked as pictured in Figure 52 for drying. Drying takes about three months, preferably in a covered area. If wood is used wet, not only will there be a marked decrease in the degree of combustion, but a significant amount of creosote, soot, and other combustion products will be deposited inside the stovepipe and weaken the draft. If these products remain in the flue, they are also liable to cause a flue fire. The frequency of such fires has been greatly exaggerated, but there is no need to have the problem at all. It can be simply and easily avoided by burning only dry wood and cleaning out the flue at the end of each heating season.

Grates should always have 1-inch spacers to keep contact between wood and grate at a minimum. Elevating the wood permits air to circulate immediately below with little obstruction from the grate itself, and burning occurs virtually all along the log. This is particularly important when a grate heat exchanger is used. Flames and burning coals at the bottom of

split logs

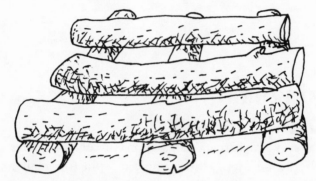

whole logs

Figure 52. Stacking Firewood

the log combine to form an area of intense heat in direct contact with the tubes, producing excellent heat transfer. There is also a reduction in smoke (indicating more complete combustion) due to the fact that a large mass of the wood no longer lies directly on the tubes and is thereby denied an adequate supply of oxygen. (One often sees the bottom of a log on a grate smoldering for long periods even as its top blazes.)

The lower tubes are also important to heat transfer because they are the entry point for room air at its lowest temperature. When this relatively cool air encounters an area of intense heat, as occurs by elevating the wood, a new and productive heat-transfer area is brought into the process. A similar situation exists at the shelf, and spacers to elevate logs serve a useful function there as well.

136

SECTION THREE
Heat Pumps

The earth's atmosphere is a vast reservoir of heat that is continuously renewed by the sun. Tapping this energy source to heat homes by means of a heat pump is not a novel concept, but one that evolved in the nineteenth century, beginning when Lord Kelvin developed a technologically sound method for "pumping heat." But there was little commercial application in that era of cheap energy, and only a few heat pumps were in use during the 1930s. By the early 1950s, central air conditioning as well as heating was becoming widespread, and it appeared that the time had arrived for a heat pump that could fulfill both purposes. Although introduced to the mass market with a good deal of publicity, heat pumps received only limited acceptance. Sales subsequently all but disappeared on account of frequent breakdowns caused by defrosting and mechanical deficiencies. A period of developmental work followed, and in the late 1960s heat pumps were again being marketed. The rising cost of fuel coupled with the performance reliability of the new models have provided a powerful impetus toward mass acceptance; in California, for example, one out of every four homes built in 1980 will be heated exclusively by a heat pump.

Figure 53 shows three types of heat-pump installations currently offered by manufacturers. The *package unit,* designed

139

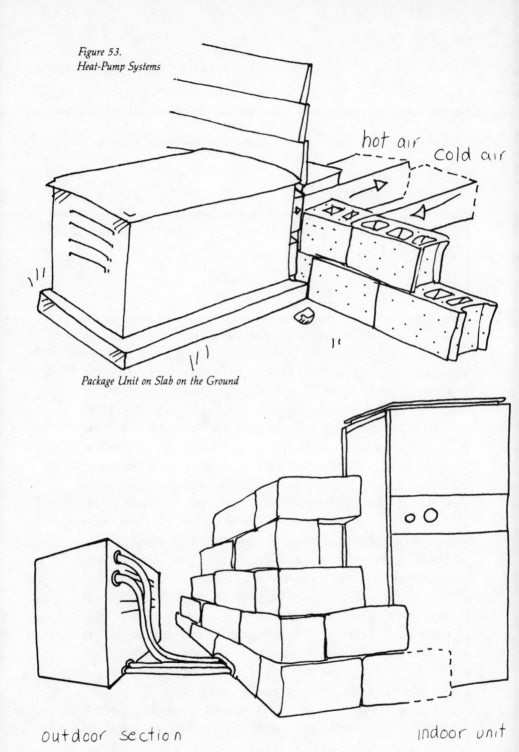

Figure 53.
Heat-Pump Systems

hot air   cold air

Package Unit on Slab on the Ground

outdoor section

indoor unit

Split Installation

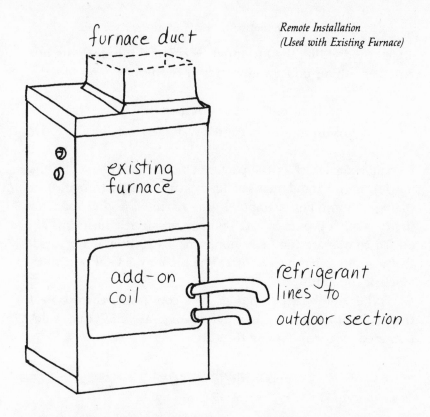

furnace duct

existing
furnace

add-on
coil

refrigerant
lines to
outdoor section

to be installed completely outdoors, is usually placed on a concrete slab on the ground or on the roof. The *split-type unit* is partly outdoors and partly indoors. The add-on type also has one part outdoors, but the other section is built onto the existing furnace, usually remote from the outdoor unit. All three types are essentially the same, though they have different applications.

Contrary to what the name suggests, a heat pump can be used either to heat or to cool a house. Although only their heating function will be described in this section, they have usually been designed to cool; therefore, when one speaks of a 5-ton unit (60,000 btus), for example, it is the cooling capacity that is rated. Manufacturers provide corresponding data for the heating capacity of the unit; the amount is always less.

When a heat pump is used to heat a house, its efficiency is rated by a *coefficient of performance (COP)*. The COP is determined by dividing the btu input per hour (btuh) into the btu output per hour under a given set of conditions:

$$\frac{\text{btuh output}}{\text{btuh input}} \quad \text{or} \quad \frac{\text{btuh output}}{\text{unit's wattage} \times 3.413}$$

Electric heat, which is 100 percent efficient (as received at the house), is the standard used to measure the performance of the heat pump. For example, if one *kilowatt hour (kwh)* of electricity (which produces 3,413 btus in an electric heater) is required to operate the heat pump and 6,826 btus is obtained, the heat pump has a COP of 2. The higher the COP, the more efficient the unit.

When the heat pump is used to cool rather than to heat the house, it is rated by an *energy efficiency ratio (EER)* that is determined by the following formula:

$$\text{EER} = \frac{\text{btuh removed}}{\text{unit's wattage}}$$

Heat pumps are generally a bit less efficient than central air conditioners.

A heat pump's cost does not necessarily reflect its efficiency or durability; prices vary according to manufacturer, capacity, and installation costs. A completely installed, 5-ton unit, including ductwork, is about $2,500 in California, but the identical installation may be higher or lower elsewhere. However, the figure of $2,500 is a good estimate of what one is likely to pay for a 5-ton unit, a capacity that fulfills the heating requirements of a 2,000-square-foot house in northern California.

The purpose of a heat pump is to extract heat from the atmosphere that can then be used to heat the house (or extract the heat from a house for cooling). Figure 54 shows how a typical heat pump operates. A fan draws a large volume of

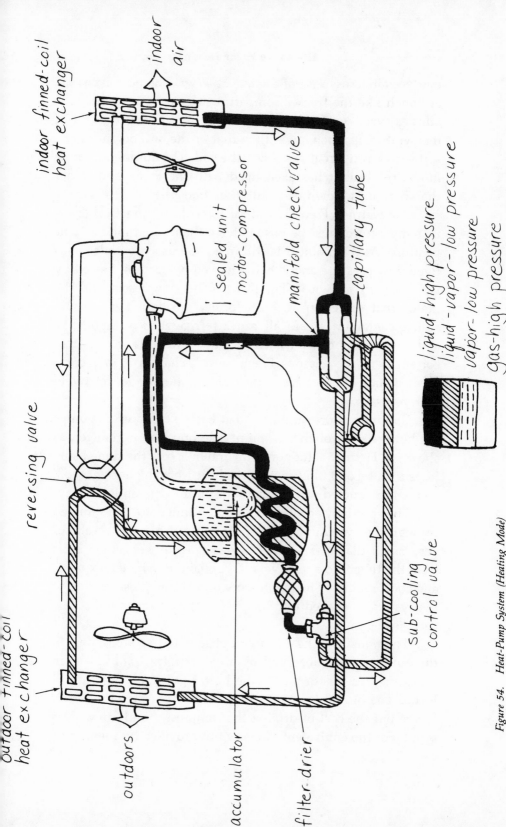

Figure 54. Heat-Pump System (Heating Mode)

outdoor air across a *finned-coil heat exchanger*. The heat exchanger, much like the freezer compartment of a refrigerator, contains a very cold refrigerant, *Freon 22*. Since the Freon temperature is much lower than that of the outdoor air, it absorbs heat as it circulates inside the coil. The outdoor air, now minus the btus it has exchanged with the Freon and colder than before, is blown back into the atmosphere by the fan.

The warmed Freon is still far too cold for practical use— its temperature must be raised above the desired indoor temperature. At this point it is still in a gaseous state under relatively low pressure and is being moved by suction created by the *compressor* shown in Figure 54. (The compressor and the motor that drives it are both located inside a hermetically sealed container, as are all other refrigeration components, and never make contact with the atmosphere.) The compressor sucks in the gaseous Freon and discharges it in a compressed state as a hot gas-and-liquid mixture under relatively high pressure.

Continuous pressure provided by the compressor pumps the hot Freon through the coil of a second heat exchanger, as shown in Figure 54. Indoor air is drawn across the finned surfaces by a second blower, absorbs heat from the hotter Freon, and is then ducted to various rooms in the conventional way.

The partially cooled Freon continues to be pumped through the system under high pressure and enters a tube of very small diameter (known as a *capillary*). The Freon exits from the tube into a relatively large chamber where it expands and cools rapidly. The Freon, now on the low pressure side of the system, is sucked into the first heat exchanger and a new cycle is begun.

The transfer of heat is accomplished in two steps; heat in the outdoor air is absorbed by a colder refrigerant and then transferred to indoor air by the hotter gas-liquid mixture of Freon. The heat that has been absorbed from the outdoor air is free but the cost of driving the compressor and fan motors is not. For the equivalent of every btu provided by the motors during each cycle, an average of 2.5 btus is recovered at a 47-degree outdoor temperature. The heat pump extracts more heat than it puts into the system, with the difference being provided by the sun.

A major factor in all heat-transfer operations is the difference in temperature between the donor of heat and the recipient. The greater the temperature difference between the atmosphere (donor) and Freon (recipient), the better the heat transfer. The heat pump is most efficient when the outdoor temperature is high; its efficiency diminishes as temperature falls.

Table 5

COP AT DIFFERENT TEMPERATURES
FOR A TYPICAL HEAT PUMP

Outdoor Temperature (Degrees Fahrenheit)	Total Watts	COP
65	3,980	2.88
60	3,840	2.79
55	3,710	2.69
50	3,570	2.60
47*	3,400	2.50
40	3,280	2.39
35	3,175	2.25
30	3,060	2.12
25	2,945	1.99
15	2,750	1.71
10	2,675	1.54

* Temperature standard used in determining COP rating of all heat pumps.

Table 5 shows the performance of an average heat pump at different temperatures. Since the COP at 47 degrees is the standard used for rating (2.5 in this case), that figure will appear on the nameplate. However, wintertime temperatures in most areas of the country are a good deal colder than 47 degrees. The heat pump can therefore be expected to operate well below its rated COP most of the time. It is misleading to imply, as current advertising does, that the homeowner who installs a heat pump will cut fuel bills by 150 percent. Even in those rare localities where the heat pump operates at an average COP of 2.5, there is no such reduction in fuel costs (unless one assumes that the homeowner had previously been using electric heat, the most expensive by far of traditional heat sources).

The COP of a heat pump is a valuable item of information for comparing the relative merits of different brands and models, but *it does not indicate the efficiency that will occur in any specific situation*. For example, a heat pump in Hawaii operating at a COP of 2.8 yields 2,077 btus per penny, but the same unit operating in Alaska at a COP of 1.8 yields only 1,335 btus per penny. The fact that heat pumps operate more efficiently at higher outdoor temperatures indicates that they are best suited for regions with mild winters. Their performance must be significantly improved for them to become viable in northerly locales.

In all climates, the cost-effectiveness of the heat pump is further reduced because an *auxiliary heat source* is needed at those times when the heat pump alone cannot fulfill demand. The problem is minor in the South, since a heat pump operating at a high COP rarely requires help from the auxiliary source. But in more severe climates, where the heat pump operates at a reduced COP, *one-third to one-half of all heat used over an average winter must be supplied by the auxiliary source.*

The problem is common to virtually all heat pumps now in use. Their designers faced the dilemma of either developing units large enough to fulfill all heating needs but therefore costly and inefficient a good deal of the time, or producing smaller, inexpensive units operating at a high COP that require supplemental heat. Manufacturers chose the latter option, sizing units to fulfill cooling requirements of the house and leaving homeowners to bear the added cost of auxiliary heating.

Almost all heat pumps employ electrical resistance *strip*

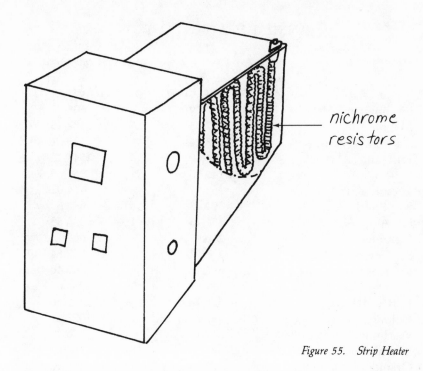

nichrome
resistors

Figure 55.   Strip Heater

heaters as the auxiliary heat source; Figure 55 shows a typical assembly. Electricity is conducted through coiled *nichrome* wires and resistance to its flow creates heat. The heat-pump fan draws air across the heated wires and the heated air is delivered to the house. (The operation is very similar to that of the ordinary electric heater.) The heat produced by the strip heater yields 741 btus per penny as opposed to 1,852 btus supplied by a heat pump operating at a COP of 2.5.

Table 6 summarizes the results of a heat-pump cost study conducted in various cities over several years. The figures show that the heat pump is modestly more cost-effective than oil in southern regions and substantially less cost-effective in northern ones; in areas where the average temperature in winter is approximately 38 degrees, the cost of operating a heat pump is about the same as for oil. Natural gas is by far the cheapest fuel, followed by oil, propane, and, finally, electricity.

Table 6

ANNUAL HEATING COSTS

Location	Heat Pump @ $.038/KWH	Conventional Elec. Heat @ $.038/KWH
Birmingham, Ala.	257.79	598.90
Mobile, Ala.	157.58	382.75
Anchorage, Alas.	1273.24	2123.31
Fairbanks, Alas.	2201.88	2541.32
Phoenix, Ariz.	145.43	359.29
Tucson, Ariz.	168.72	414.02
Fort Smith, Ark.	299.45	684.63
Los Angeles, Cal.	152.43	383.88
Sacramento, Cal.	266.97	660.19
San Francisco, Cal.	290.21	727.84
Colo. Springs, Colo.	692.78	1360.40
Denver, Colo.	637.98	1265.34
Hartford, Conn.	658.76	1306.66
Wilmington, Del.	467.50	1023.22
Washington, D.C.	387.88	876.79
Miami, Fla.	24.83	62.40
Orlando, Fla.	78.84	195.42
Tallahassee, Fla.	152.68	370.11
Atlanta, Ga.	262.66	616.98
Savannah, Ga.	179.77	432.01
Boise, Idaho	589.95	1254.23
Chicago, Ill.	720.02	1360.01
Springfield, Ill.	474.00	1016.74
Evansville, Ind.	453.00	965.78
South Bend, Ind.	693.34	1338.46
Des Moines, Iowa	746.99	1352.91
Sioux City, Iowa	822.55	1438.59
Topeka, Kans.	555.61	1099.38
Wichita, Kans.	487.11	997.54
Lexington, Ky.	490.69	1019.14
Louisville, Ky.	431.36	930.17
New Orleans, La.	122.63	303.13
Shreveport, La.	207.59	496.59
Portland, Ma.	773.24	1491.30
Baltimore, Md.	441.89	970.96
Boston, Mass.	547.06	1158.72
Detroit, Mich.	641.64	1286.43
Flint, Mich.	796.98	1484.52
Duluth, Minn.	1269.89	1990.82
Minneapolis, Minn.	976.86	1613.01
Jackson, Miss.	218.37	515.87

Table 6 (cont'd)

ANNUAL HEATING COSTS

Oil Heat @ $.461/Gal.	Propane Heat @ $.301/Gal.	Nat. Gas Heat* @ $.207/Therm†
254.49	265.46	161.32
162.64	169.65	103.10
902.27	941.13	571.95
1079.90	1126.41	684.55
152.68	159.24	96.78
175.93	183.51	111.52
290.93	303.46	184.42
163.12	170.15	103.41
280.54	292.62	177.83
309.28	322.60	196.06
578.08	602.98	366.45
537.69	560.85	340.84
555.25	579.16	351.97
434.80	453.53	275.62
372.58	388.63	236.18
26.52	27.66	16.81
83.04	86.62	52.64
157.27	164.05	99.70
262.18	273.47	166.20
183.58	191.48	116.37
532.97	555.93	337.85
577.92	602.81	366.34
432.05	450.66	273.88
410.39	428.07	260.15
568.76	593.26	360.54
574.90	599.66	364.43
611.31	637.64	387.51
467.17	487.29	296.14
423.89	442.15	268.71
433.07	451.72	274.52
395.26	412.29	250.56
128.81	134.36	81.65
211.02	220.11	133.77
633.71	661.00	401.71
412.60	430.37	261.55
492.38	513.59	312.12
546.65	570.19	346.52
630.83	658.00	399.88
845.97	882.41	536.26
685.42	714.95	434.49
219.21	228.66	138.96

Table 6 (cont'd)

ANNUAL HEATING COSTS

Location	Heat Pump @ $.038/KWH	Conventional Elec. Heat @ $.038/KWH
Kansas City, Mo.	495.68	1003.96
Springfield, Mo.	468.56	978.10
Great Falls, Mont.	917.47	1584.77
Omaha, Neb.	696.37	1282.08
Las Vegas, Nev.	221.26	537.71
Reno, Nev.	642.71	1327.44
Newark, N.J.	468.06	1023.83
Albuquerque, N.M.	406.03	906.06
Binghampton, N.Y.	816.33	1519.33
New York, N.Y.	441.00	974.96
Charlotte, N.C.	293.31	682.95
Greensboro, N.C.	368.43	825.50
Bismark, N.D.	1220.59	1849.40
Cincinnati, Ohio	505.72	1051.61
Youngstown, Ohio	668.63	1322.96
Oklahoma City, Okla.	374.01	816.30
Tulsa, Okla.	354.69	777.43
Medford, Ore.	455.57	1073.81
Portland, Ore.	422.85	1015.07
Philadelphia, Pa.	451.32	993.30
Scranton, Pa.	678.70	1336.06
Providence, R.I.	570.50	1199.39
Charleston, S.C.	200.65	480.31
Columbia, S.C.	250.03	585.03
Huron, S.D.	1042.93	1674.44
Rapid City, S.D.	853.41	1522.17
Memphis, Tenn.	294.21	676.96
Nashville, Tenn.	354.83	784.79
Amarillo, Tex.	440.21	936.84
Austin, Tex.	164.06	396.18
Brownsville, Tex.	59.23	147.32
Salt Lake City, Utah	585.13	1219.24
Burlington, Vt.	907.23	1593.11
Norfolk, Va.	301.97	708.47
Roanoke, Va.	388.32	876.93
Seattle-Tacoma, Wash.	474.14	1138.14
Spokane, Wash.	682.28	1418.74
Charleston, W.Va.	465.20	980.50
Green Bay, Wisc.	968.58	1677.93
Madison, Wisc.	842.20	1508.05
Casper, Wyo.	834.47	1555.85

* Furnace rated at 70% efficiency.
†1 Therm = approximately 100 cubic feet.

Table 6 (cont'd)

ANNUAL HEATING COSTS

Oil Heat @ $.461/Gal.	Propane Heat @ $.301/Gal.	Nat. Gas Heat* @ $.207/Therm†
426.62	444.99	270.43
415.63	433.53	263.47
673.42	702.42	426.88
544.80	568.27	345.35
228.49	238.33	144.84
564.08	588.37	357.57
435.06	453.80	275.79
385.02	401.60	244.06
645.62	673.43	409.26
414.30	432.14	262.62
290.21	302.71	183.96
350.78	365.89	222.36
785.88	819.73	498.17
446.87	466.11	283.27
562.17	586.39	356.36
346.88	361.82	219.89
330.36	344.59	209.42
456.30	475.96	289.25
431.34	449.92	273.43
422.09	440.27	267.56
567.74	592.19	359.89
509.66	531.62	323.08
204.10	212.89	129.38
248.60	259.31	157.59
711.53	742.18	451.04
646.82	674.68	410.02
287.66	300.05	182.35
333.49	347.85	211.40
398.10	415.24	252.35
168.35	175.60	106.72
62.60	65.30	39.68
518.10	540.42	328.43
676.97	706.13	429.13
301.06	314.02	190.84
372.64	388.69	236.22
483.64	504.47	306.58
602.87	628.84	382.16
416.65	434.60	264.12
713.01	743.73	451.98
640.83	668.43	406.22
661.14	689.61	419.10

Manufacturers are aware of the unfavorable cost picture of the heat pump in most of the country and are developing units that are more efficient at lower outdoor temperatures and also eliminate supplementary electric heating.

Figure 53 (see page 141) shows a heat-pump heat exchanger mounted on an oil-fired hot-air furnace so that the two systems can be used together. The exchanger is the same as those used in split or packaged units and is served by the furnace blower. Except for longer refrigerant lines, the arrangement is identical to the ordinary installation.

The most significant advantage of this "remote" system is that the heat pump is employed only when it is economical to do so. For example, if use of the heat pump is limited to periods when the outdoor temperature is 40 degrees or higher, it will operate at a relatively high COP (actually providing 100 btus more per penny spent compared to the exclusive use of the furnace). Furthermore, cost-effectiveness of the oil furnace will also be higher when used in conjunction with the heat pump than when used exclusively to provide a home's heating needs.

Home oil furnaces are usually rated at an efficiency of 65 percent, but under normal operating conditions only 50 to 55 percent efficiency is obtained. The discrepancy is mainly due to the poor performance of the furnace during the "warm-up" periods that occur each time the furnace comes on. By employing the furnace only half or two-thirds of the time—and under optimum conditions when outdoor temperature falls below 40 degrees—the overall number of warm-up periods is significantly reduced, thereby raising efficiency to 65 percent or higher. The result is a net gain of 200 btus per penny spent. The heat pump used in conjunction with an oil furnace costs 15 percent less to operate in the Northeast than a furnace alone; cost-effectiveness of heat pumps increases the less often auxiliary heat is used.

Let us assume that your home now has an oil furnace and

you want to add a heat pump. Since virtually all the ductwork is already in place, the cost should be no more than $2,000; this amount would be amortized in a 1,500-square-foot house within about fifteen years. Even though the cost of heating your home will continue to rise on account of increasing prices for electricity, the greater efficiency of the heat-pump–oil-furnace installation will lower overall heating costs. If the price of fossil fuels continues to rise at the same rate in the future as it has in the past, amortization time will actually be reduced to eight to ten years.

In order to determine at what temperature the heat pump becomes more efficient than the auxiliary heating system, you must know your home's *thermal balance point*. This is the specific outdoor temperature at which the increasing loss of heat in a particular house equals the declining output of the heat pump. Above this temperature, the heat pump can supply all the heating needs of the house and the use of more costly supplementary heat is avoided. The balance point generally occurs between 35 and 28 degrees. As a standard practice in every installation, heat-pump installers determine the balance-point outdoor temperature by collecting heat-loss data specific to each home (glass area, orientation, infiltration, insulation, etc.) and matching the information to factory-supplied data.

All heat pumps use a *two-stage thermostat* (Figure 56) to regulate the flow of heat to the house. The thermostat is mounted on an interior wall away from drafts, and the user simply sets the desired temperature in the usual way. When heat is required, a sealed bulb straightens and the mercury it contains forms an electrical path between the wires attached at both ends, thereby activating the heat pump. If the outdoor temperature falls below the thermal balance point, the heat pump will be unable to fulfill the demand for heat, and the temperature indoors will continue to fall. Another bulb then straightens to activate the auxiliary heat source. When the demand for heat is satisfied, the bulb tilts and mercury collects at one end, breaking the electrical path and shutting off the heating system.

Two-stage thermostats are adequate in mild climates but perform poorly in regions where temperatures fall below freezing. Lacking any balance-point control, they often turn on the more costly auxiliary system when there is no valid reason for doing so. For example, if the balance point of a particular installation is 30 degrees and the outdoor temperature is 35 degrees, it is unnecessary to activate the auxiliary system; the heat pump will provide enough heat even though it will take relatively longer to do so because it has a limited capacity and is also operating at a reduced COP. In this situation, one often raises the thermostat setting above its normal position to speed up the process. A similar situation occurs in the morning as the thermostat setting is raised from its lower nighttime position and the heat pump cannot fulfill the sudden demand. A door or window left open inadvertently will also engage the auxiliary heat source unnecessarily and lower cost-effectiveness of the system.

Many manufacturers realize the need for thermal balance-point control and offer an optional outdoor thermostat as an

Figure 56.   Two-Stage Thermostat

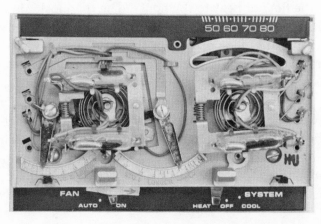

accessory in addition to the two-stage thermostat. If the balance point of a particular installation is determined to be 30 degrees, for example, the outdoor thermostat is set there. As long as the outdoor temperature remains 30 degrees or higher, auxiliary heat cannot be used and the system functions solely on the heat pump regardless of indoor temperature. When the temperature falls below 30 degrees, the heat pump is automatically shut off and the auxiliary heat turned on.

An outdoor thermostat is an improvement over the use of the indoor one alone but has a number of defects that cause its performance to be less than satisfactory. Outdoor thermostats do not have accurate or reliable sensing mechanisms. False readings often occur and cause misuse of the two heating stages. For example, if the thermal balance point is 30 degrees and the outdoor temperature is 37 degrees but there is snow on the ground, the thermostat will sense a temperature of 29 degrees and mistakenly activate the auxiliary system. Conversely, if the outdoor temperature is 24 degrees but the sun strikes the thermostat, a temperature of 40 degrees will be "sensed" and the heat pump will be activated even though the auxiliary system should be used. Many similar conditions "fool" the outdoor thermostat, thereby making it perform unsatisfactorily as the means of balance-point control.

Another problem common to heat pumps is that the cold outdoor coil attracts not only heat in the atmosphere but moisture as well; this moisture condenses on the coil in the same way that frost collects in the freezer compartment of a refrigerator. If the outdoor temperature is near or below freezing, condensed moisture will freeze on the coil. The greater the buildup of ice, the less efficient the heat pump, because the ice "insulates" the coil and reduces the amount of heat extracted from the air.

Although moisture is always present in outdoor air, the amount varies according to the weather. Moisture content is highest during periods of winter snow, sleet, and fog. During these times, the need for *defrosting* is greatest. In drier weather defrosting is required less often and at times can be omitted entirely. Defrosting is an essential part of heat-pump operation and the method utilized plays a significant part in cost-effectiveness. In all but a few advanced heat pumps, manufacturers have designed automatic defrosting cycles that operate on a *time-temperature* basis. A timer preset at the factory activates the defrost cycle every 30, 45, 60, or 90 minutes. The outdoor temperature used in conjunction with the time interval is usually 45 degrees; frost does not ordinarily form at higher temperatures. At the preset time and temperature, a reversing valve switches the heat pump from the heating mode to its cooling mode. Then, with the heat pump used as an air conditioner, the outdoor coil, acting as a condenser rather than evaporator, receives the hot liquid-gas mixture of Freon that melts the accumulated ice.

Although heat pumps should be defrosted only when there is a sufficient buildup of ice, in a time-temperature cycle hot Freon is pumped to the outdoor coil even if there is no ice on it; *defrosting occurs whether it is needed or not*, reducing both the efficiency of the heat pump and its service life expectancy. During defrosting, auxiliary heat is turned on automatically to eliminate cold drafts produced when the system is switched to the cooling mode. The waste of heat accompanying each cycle is small, but over the course of the year thousands of needless cycles significantly decrease the heat pump's cost-effectiveness. Additionally, the extreme heat and flood-backs accompanying each shift of the defrost cycle are particularly hard on the sealed unit. Such unnecessary wear and tear on the machine contributes to premature breakdowns. Thus, one must conclude that the time-temperature basis for defrosting is a poor and unacceptable arrangement; homeowners should avoid heat pumps that still use this outmoded mechanism.

The problems of control associated with both defrosting and thermal balance point cannot be passed off as "the state of the art." Solid-state technology produced solutions years ago, with control devices far more accurate and reliable than their electro-mechanical counterparts. Ironically, some of the companies that manufacture heat pumps also pioneered solid-state development; but while such technological advances are being widely applied to automobiles, television sets, and many other products, aside from a few notable exceptions they have not been applied to the heat pump.

A few companies, recognizing the need for better controls, have abandoned the time-temperature basis for defrosting and have introduced a solid-state resistant device for monitoring and controlling the cycle. The device is built into the heat pump and measures the temperature inside the coil within two degrees of accuracy. As ice builds up on the coil, its internal temperature falls; when the temperature drops to the level signifying a need for defrosting, the reversing valve automatically switches the heat pump from heating to cooling. This arrangement allows far more efficient operation than the time-temperature method, since it is responding to the actual condition of the coil. However, changes in the internal temperature of the coil can be brought about by conditions other than ice buildup, and the device has no means of differentiating among the various causes. Furthermore, since the device deals solely with defrosting and not with many other control needs, its usefulness is limited.

There are several other control devices that regulate a heat pump's functions whose performance would be improved if solid-state technology were applied to them. For example, since no heat pump (or air conditioner) should be restarted immediately after it has stopped (in order to allow internal pressures to equalize and to prevent damage to the compressor), a control mechanism is required for this purpose. Other controls are necessary to monitor the compressor temperature and pressure and to turn off the machine when design limits are exceeded; yet another is needed to activate the heat pump above the balance-point temperature and the aux-

Table 7

COMPARISON STUDY OF DEFROST FREQUENCY

City	Heat Pump Size in Tons	Computer-Controlled Heat Pump Defrosts:	
		# times	# hours
York, Pa.	3	200	17.35
Tacoma, Wash.	2½	154	11.2
Denver, N.C.	3½	309	18.0
Mineral Wells, Tex.	2½	179	16.5
Winder, Ga.	4	79	4.2
Ithaca, N.Y.	2½	290	16.5
Chesterfield, Mo.	4	524	42.6

iliary heat below it. In the air-conditioning mode, dehumidification also needs controlling. In most heat pumps it is performed automatically during each cycle irrespective of whether interior air is moist or dry. Since the loss of efficiency is significant, advanced heat pumps are designed to dehumidify the air only when there is an actual need to do so.

At this writing only the York Company has tackled the entire control problem by utilizing solid-state components; it has also introduced a computer-activated module that accurately and reliably controls many operations of the heat pump. For example, an anticycle delay that is built into the module prevents the unit from starting for five minutes after it has been stopped and allows internal pressures to equalize. The module, located inside the heat pump, is not subject to the problems encountered by an outdoor thermostat and controls balance-point operation effectively. It also contains an integrated solid-state timer circuit that measures air and internal temperatures before defrosting and cannot be "fooled" by sudden and temporary changes in outdoor conditions. By comparing accurate sensing data from various locations inside the system and outdoors, the module is able to evaluate when there is a real need for defrosting and then to activate the defrost mechanism.

Table 7 (cont'd)

An Ordinary Heat Pump Would Defrost During a

30-minute cycle		60-minute cycle		90-minute cycle	
# times	# hours	# times	# hours	# times	# hours
3592	179.6	1796	89.8	1197	59.8
2372	118.6	1186	59.0	790	39.5
2145	107.2	1072	53.6	715	35.7
1022	51.1	511	25.5	340	17.0
2364	118.2	1182	59.1	788	39.4
5972	298.6	2986	149.3	1990	99.5
3183	159.1	1591	79.5	1061	53.0

Table 7 presents the results of a defrosting study conducted in representative areas of the country in which heat pumps of various capacities employing the computer-controlled module were compared to similar-sized units using the time-temperature method. (The time period of the test was October 1, 1977 to April 30, 1978.) The results demonstrate the superiority of the module arrangement. In Ithaca, New York, for example, the computer-controlled heat pump defrosted only 290 times while the time-temperature heat pump on a 30-minute cycle defrosted 5,972 times. Several manufacturers are now following the direction taken by the York Company, so that other heat pumps using solid-state modules to control their operation can soon be expected on the market.

Although good controls are essential to increasing a heat pump's cost-effectiveness, they cannot raise the actual capacity of a unit. The majority of heat pumps currently on the market still are sized according to the cooling needs of the house and continue to fall short of adequately meeting heating requirements in moderate and severe climates. Add-on units are economical if there already is an existing oil furnace to supplement the heat pump, but the new-home builder considering this type of heating must spend $2,200 or so for the furnace in addition to the cost of the heat pump.

In order to eliminate the need for an expensive auxiliary heating system, a few manufacturers have designed two-speed heat pumps that are capable of fulfilling total heating requirements in all climates. The heart of every heat pump is its compressor, and the two speeds refer to the movement of its piston. When such a pump operates at the lower speed, the compressor is rated at 1,750 rpm as opposed to 3,500 rpm at the higher speed. At high speed, the heating capacity of the unit increases by about 40 percent.

The map shown in Figure 57 divides the United States into two areas demarcated by a line that represents 3,000 *heating degree days (HDD)*.* Below the line, two-speed heat pumps are sized to cooling needs and above it to heating needs. In both areas the heat pump operates at the slower speed when demand for heating does not exceed capacity, then switches automatically to high speed and greater capacity when the demand for heat cannot be fulfilled.

Although a sufficiently large, well-designed two-speed heat pump is capable of heating most homes without the need for auxiliary heating, there is one significant drawback: the two-speed heat pump is slightly less efficient than its standard counterpart when it operates at its slow compressor speed. Table 8 shows the performance of six typical two-speed units of different sizes operating at low and high speed. (TC = total net capacity in btuh; "High-Temp" heating standards call for 70-degree air entering indoor component, 47-degree air entering outdoor component; "Low-Temp" heating standards call for 70-degree air indoors, 17-degree air outdoors.)

The figures for total capacity show that the two-speed

---

* The number of heating degree days in a given day is determined by subtracting the average temperature of the day from 65 degrees. For example, if the average temperature in a 24-hour period is 40 degrees, there were 25 heating degree days that day.

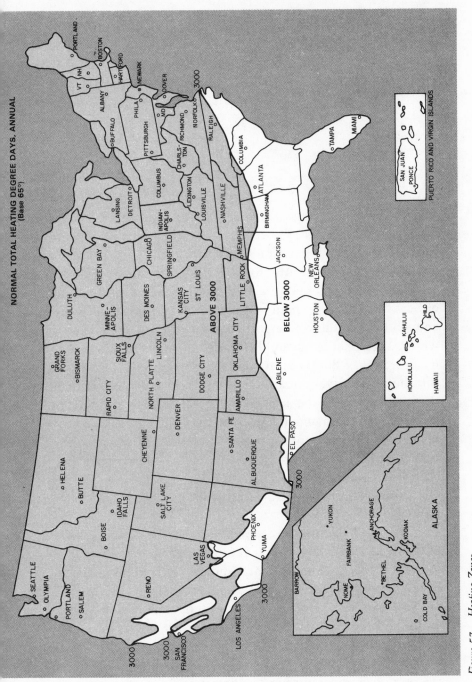

NORMAL TOTAL HEATING DEGREE DAYS, ANNUAL
(Base 65°)

Figure 57. Heating Zones
National Division by 3,000 Heating Degree Days

161

Table 8

High-Temp Heat		Low-Temp Heat	
TC	COP	TC	COP
21,500	2.9	10,500	1.8
30,500	2.8	16,500	1.9
37,500	2.8	21,000	2.1
44,000	2.8	22,500	2.0
50,000	2.8	26,500	2.0
60,000	2.8	32,000	1.8

(high-temp) heat pump delivers far more heat than do standard (low-temp) units. However, when the outdoor temperature is 17 degrees, the most efficient unit delivers 37,500 btuh, an amount that is too little to heat a 1,500-square-foot house in northerly climates. While sufficient to replace the amount of heat lost during an hour, a rate of 37,500 btuh is not nearly enough to raise the temperature of a cold house to 68 degrees within a reasonable time. Oil furnaces generally have a capacity in excess of 80,000 btuh and are therefore able to warm a cold house relatively quickly. Much larger heat pumps than those given would be needed before an equivalent performance could be obtained. However, the COP figures that correspond to the different capacities show that as the size of the heat pump increases beyond 37,500, its low-temperature efficiency decreases. A unit capable of heating a home in a northerly climate would have a COP of about 2.5. (See Figure 58 for national COP ranges.)

The COP figures also show that all six heat pumps have high efficiencies at high temperatures and compare favorably with oil furnaces or equivalent standard heat pumps similarly operating at an outdoor temperature of 47 degrees. However, at this temperature the heat pump will operate at slow speed, so its high-speed efficiency is irrelevant. It is the low-temperature efficiency that reveals what will occur under actual operating conditions. All units have a COP around 2.0 when the outdoor temperature is 17 degrees and the compressor is operating at high speed.

A COP of 2.0 indicates that the two-speed pump performs a good deal better than a standard heat pump operating under the same conditions; it also reflects a marked improvement of efficiency in cold climates where heat demand is greatest—typically the most adverse conditions for a heat pump. And since there is no need for an oil furnace, the initial outlay is much less (and of course, the system provides air conditioning as well). Consequently, the two-speed heat pump has been acknowledged to be a major innovation to which more and more manufacturers are turning.

When it operates at a COP of 2.0, the two-speed heat pump provides 1,482 btus per penny compared to 1,978 btus per penny for oil; so at this COP it is a good deal less cost-effective. As the outdoor temperature rises above 17 degrees, the COP increases correspondingly and the picture changes for the better. But even at an average operating temperature of 32 degrees, the COP of the heat pump is only 2.3; in order for the pump to become competitive with oil, its COP must

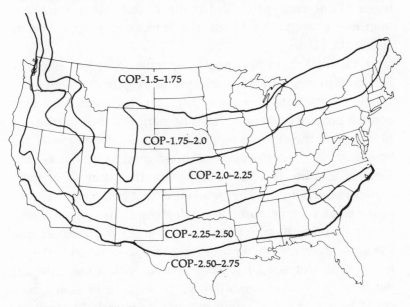

Figure 58. Approximate Annual Ranges of Heat Pump COP (Including Supplementary Strip Heater)

rise to 2.8. In northerly climates, most of the fuel consumed during the winter occurs when outdoor temperatures are 35 degrees or less; since the heat pump will then be operating at a reduced efficiency, *it will cost the homeowner 30 percent more to use the heat pump rather than an oil furnace.*

The picture is not likely to change as the price of oil continues to rise because the cost of electricity will most likely rise as well. Of course if the local cost of electricity is below the national average of 4.6¢ per kwh, cost-effectiveness will be better; if the price is 3.2¢ per kwh or lower, the two-speed heat pump becomes the logical choice as the sole heating system. However, since most homeowners do not live in areas of cheap electricity, the use of a two-speed heat pump will remain too expensive until manufacturers develop models whose COP is at least 2.8 when the outdoor temperature is below freezing.

Although the problem of increasing efficiency at low temperatures has not yet been satisfactorily solved, manufacturers of the two-speed heat pump have made a number of improvements that hold forth the promise of greater efficiencies in the future.

When standard heat pumps operate under low-temperature conditions, they are subjected to very high stresses that often cause serious damage to the unit. Most standard units contain protective devices that automatically shut off the heat pump when the outdoor temperature falls below a preset safe operating limit. (Zero degrees is often used as the lowest safe temperature, but there are many units that shut off at much higher temperatures.) The two-speed heat pump can be safely operated at a temperature of -25 degrees. It also defrosts safely at an outdoor temperature of 30 degrees, much lower than the 45 degrees for standard units, and also eliminates the extreme heat and flood-backs associated with going into and out of a cycle. Most importantly, the amount of heat a two-speed heat pump can safely produce—particularly when the outdoor temperature is at or below freezing—is much greater. These and similar advances, combined with improvements still in the laboratory or pilot stage of development, will undoubtedly lead to vastly improved performance. Nevertheless it will take several more years before a heat pump is produced that is viable for homeowners who live above the 3,000-HDD line in Figure 57.

If you live below the line and are planning to build a home, a heat pump is your best buy if wood isn't a feasible alternate fuel. The heat pump is less costly to install and maintain than the combination of an oil furnace and central air conditioning; it occupies a relatively small space, and can be located entirely outdoors; and it retains the convenience of thermostatic control. Ideally, it should have a two-speed compressor, solid-state controls, and an economizer that permits adjusting the dehumidification operation to your specific needs. A heat pump with these features will cost about the same to operate as a system combining an oil burner with central air conditioning.

If you already have an oil furnace but want air conditioning, it is still more cost-effective to install central air conditioning than a heat pump.

If you are presently using electric heat and pay 4.6¢ per kwh or more, switching to a heat pump is advisable; its cost will be amortized in about ten years by the savings in electricity.

If you live above the 3,000-HDD line, the best you can hope for is that the cost of the heat pump will equal the use of oil; but most likely the heat pump will be a good deal more expensive.

Before deciding to use a heat pump, you should consider the fact that it is a machine that indirectly consumes fossil fuels, since it receives its energy from the oil, natural gas, and coal used by utilities to generate electricity. The rising cost of fossil fuels leads to a corresponding rise in cost of electricity, so the operation of the heat pump will become more expensive. The situation will cause an increasing burden if we regard the heat pump as the sole source of heat for homes. However, if we regard the heat pump solely as an *auxiliary* heat source and use it only when most advantageous, a dramatic reduction in the cost of heating a home is possible.

The sophistication of heat pumps is best used to supply heat temporarily in the early morning, or other such conveniences, and not for overall heating. If we limit heat-pump use to 10 percent, for example, and use a combination of wood and solar energy for the remainder of our needs, the yearly fuel cost for a 1,500-square-foot house is reduced by $400 or more, and the comfort of automatic control is retained. This type of heat-pump supplementary use is economically sound for all but the coldest climates and is a viable choice for the homeowner.

Afterword: Utilizing Heat Efficiently

Increasing the efficiency of any heat-producing system is of little value if the heat produced is quickly lost or accumulates in unwanted locations. Since the advent of the energy crisis, an advertising blitz has been directed at increasing insulation in homes, to the point where the idea of minimizing heat loss has become synonymous with insulation behind walls, ceilings, and floors. Well-intentioned legislators have passed laws granting tax incentives to homeowners who insulate, and municipal agencies have increasingly incorporated the practice into building codes.

The value of insulation has been known for a long time to builders, who have installed it in new or renovated homes as a regular practice for the past fifty years. The vast majority of the sixty million single-family homes in the country today were adequately insulated long before the energy crisis. Fiberglass sandwiched between paper barriers has been the predominant material, though since the 1950s aluminum foil has been substituted for the paper facing the room. In standard practice, a batt of $2\frac{1}{2}$-inch-thick fiberglass is stapled into bays, leaving a 1-inch dead-air space between the back of the insulation and the sheathing. The arrangement has been effective, but about fifteen years ago manufacturers stopped producing insulation in $2\frac{1}{2}$-inch thicknesses and increased the minimum to $3\frac{1}{2}$ inches. An immediate 30 percent increase in sales resulted, although it is highly debatable whether $3\frac{1}{2}$ inches is more effective than $2\frac{1}{2}$ inches combined with a dead-air space. In any case, the difference between the two arrangements is negligible, and the owner of a new home has been saddled with an additional cost for little value. (The average new home of today is much tighter than older ones, due to the use of sheathing and felt under exterior siding; air passage through walls and the convection inside them has been greatly curtailed, and heat loss diminished accordingly.) Fiberglass insulation $2\frac{1}{2}$ inches thick in walls and between rafters, $3\frac{1}{2}$ inches between ceiling joists, and none between floor joists for houses with basements or crawl spaces is a cost-effective installation.

Insulating a house decreases heat losses by 17 to 20 percent, so I would hardly suggest that the practice of insulating be abandoned. But we are dealing today with the *excessive* use of insulation, beyond appropriate amounts and cost-effectiveness. Infrared photographs of houses showing heat losses as orange areas confirm the fact that heat retention does not depend on more and more insulation.

Glass is the major avenue of heat loss in a house, and until the main thrust of energy-saving is directed to it, no major improvement is likely. Building codes require that at least 10 percent of the square footage of a room be allocated to sash (the movable part of a window), but most homes have a good deal more such area and also large expanses of fixed glass that cause substantial heat losses. Glass accounts for 55 to 70 percent of a house's total heat loss, with the higher figure more in line with today's modern designs.

Comparative tests of various glass arrangements have shown that an ordinary single pane served by an insulated shutter is by far the most efficient and cost-effective way of utilizing and retaining heat in a window. (Alternatives such as thermal blinds, thermal draperies, thermal window shades, etc., are not only less efficient but also far less cost-effective.)

Due to their high cost, wood-frame windows have been largely supplanted in new homes by aluminum and steel. The metals provide a ready path for heat loss by conduction, and the initial saving is soon used up by higher fuel costs. Wood-frame windows remain the most cost-effective over the long run.

Infiltration—cold air admitted to the house through a variety of openings, and warm air lost through them—is another significant way heat is lost from a house. It is just as important a factor as insulation, and infrared photographs have shown that infiltration heat losses occur mainly around exterior doors and windows. (They do *not* occur in significant amounts at foundation walls; the photos confirm that insulation under floors of houses with crawl space or basements is a waste.)

Infiltration gaps around doors have three causes: wood contraction during winter, a poor fit between the bottom of the door and saddle, and intermittent contact between the door and stop caused by warpage or poor mounting. Many weatherstripping products—adhesive-backed foam rubber, felt and metal, spring brass, and vinyl inserts—are advertised as the solution, but all are unsatisfactory. The only effective weatherstripping for doors is matched V-metal, a type widely used in better homes before World War II and rarely since. In this installation the perimeter of the door is routed to receive a small V-shaped stainless-steel strip, and a hollow V-shaped matching strip is attached to the frame. An interlocking saddle is used to close the gap below the bottom of the door. Subsequent "movement" of the door is controlled by the clamping action of the interlocking metal strips, and gaps are eliminated in all weather. This type of weatherstripping is highly efficient and durable; in New York City, for example, such weatherstripped doors have remained effective for a century.

Infiltration also occurs around the perimeter of glass due to the absence or breakdown of putty, and around sash that is poorly fitted (as evidenced by rattling). All lumberyards carry interlocking weatherstripping for loose sash and, of course, glazier's putty for resealing panes.

A less obvious infiltration occurs around the perimeter of the window frame from air movement through hollow areas left during wall framing. Some builders fill the cavities around the window frame with loose fiberglass, which will eliminate such drafts, but many simply cover them with trim. If one runs a finger along the joint between wall and interior trim and feels a temperature difference, the trim should be removed and cavities filled with fiberglass. Before putting the trim back, parallel beads of silicone caulking should be laid on the edge of the frame and wall to overcome wall irregularities and seal the trim all along its length. One should also lay a bead of caulking between the bottom of the window sill and the exterior siding, a location that is often ignored when builders are not overly concerned with heat retention.

Heat is lost each time an exterior door is opened, a problem for which vestibules used to be the answer. Vestibules have disappeared for the most part in modern home construction, but they are still the most effective way of dealing with exterior-door heat loss. One cannot hope to make a vestibule cost-effective if fuel savings alone are considered, but like a greenhouse, a vestibule can serve many functions—in addition to being a depository for wet clothing, muddy boots, umbrellas, etc.—and be a worthwhile addition to a home's living space.

Oil furnaces provide heat for approximately 80 percent of the nation's homes and despite the increasing price of oil, will continue to be the major source of heat in the coming decade. Some people have found comfort in the mistaken belief that at least the furnace is a highly efficient way of heating, and the 65 to 70 percent efficiencies on the nameplates of manufacturers would seem to substantiate this belief. However, nameplate efficiency is true only in the factory; efficiency in the home is usually around 50 percent. The lower efficiency is due mainly to the practice of almost invariably installing a much larger furnace than needed. The oversized capacity is convenient because it enables heating demands to be filled very rapidly. When one returns to a frigid house, say, the temperature can be raised to 70 degrees within ten minutes or so. But this situation is unusual; most often the furnace is simply called upon to replace normal heat loss—from 5,000 to 15,000 btuh. Except for very windy days with ambient temperatures below zero degrees, heating needs would be fulfilled satisfactorily by a furnace with less than half the capacity of the one that is usually installed.

When a home has a needlessly large furnace, a significant proportion of the heat generated in the start-up period is wasted in overheating the ducts, firebox walls, etc., thereby drastically reducing overall efficiency. The situation is analagous to driving a mile at 100 mph rather than at 50 mph; at the higher speed you arrive at your destination more quickly, but driving more slowly fuel efficiency is much higher. By increasing the proportion of heat produced *after* the start-up period (and reducing the number of times the furnace turns on in a given period), you can get closer to the efficiency on the nameplate.

A simple and inexpensive way to readjust the capacity of an oversized furnace is to reduce the size of the nozzle that delivers oil to the firebox. The capacity can generally be reduced by more than half without adversely effecting performance, but the optimum reduction in capacity varies with each particular home. A heating contractor should be consulted so that the change is tailored to meet the particular needs of the specific house. The use of a smaller nozzle can increase efficiency of the oil furnace by at least 10 percent and is highly recommended.

The water heater is another significant source of heat loss. Approximately 17 percent of the energy used in a house goes into the production of hot water. Most of this energy is consumed in maintaining the desired temperature, and a smaller amount in raising the initial temperature of the water. A thermostat controls the temperature, ordinarily set at 145 degrees. Whoever installed the water heater left the thermostat at that setting because dishwasher manufacturers recommend a temperature of 140 degrees (allowing a drop of 5 degrees between heater and appliance). The reason manufacturers recommend 140-to-145-degree hot water is because water that is less hot cannot break down animal fats left on kitchenware, and of course manufacturers want their products to perform well. But 140 degrees is far too high for all other household uses; 120 degrees is more than adequate. If the thermostat is reset at 125 degrees, one must be prepared to scrape away most of the animal fats from dishes before they go into the dishwasher, but this small inconvenience can cut hot-water costs approximately 30 percent, or $100 yearly for a four-person household.

The efficient production and retention of heat in a house is minimized if hot air accumulates at the ceiling, around the furnace, and in the attic—all areas unlikely to warm a house's occupants. Yet this is what happens in most homes, because the design of the oil-fired furnace system emerged in the cheap-energy era and no attention was paid to utilizing the heat after it was produced. Recirculation of heat that accumulates in unwanted areas is not part of a duct-and-furnace-fan system, and improved utilization requires supplemental help. The best response is a thermostatically controlled combination of ducts, registers, and a blower to circulate heated air from useless to useful areas. Minimal response is a fan mounted near the ceiling to direct heated air downward, and registers in the floor above the furnace area.

Some of the many other ways to better utilize heat include: balancing heat delivery by means of dampers in ducts, so that more heat is sent to colder rooms and less to warmer rooms; keeping the doorway to the second floor closed during the day and opening it an hour or so before bedtime; shifting furniture away from heat sources; and transferring heat from the clothes-dryer exhaust.

Some of the many ways to reduce energy costs include: electronic ignition for gas stoves to eliminate the pilot light; the use of "cool light" fluorescent tubes in place of incandescent bulbs; a thermostat timer for the furnace that automatically reduces the temperature at night; a hot-water heater under the kitchen sink that delivers instant boiling water; and, of course, switching off lights and appliances when they are not in use.

Index